GW01452997

KEEP (

KEEP ON TRUCKIN'

40 Years on the Road

Mick Rennison

Old Pond

PUBLISHING

First published 2016

Copyright © Mick Rennison 2016

All rights reserved. No part of this publication may be reproduced, stored in a retrieval system, or transmitted, in any form or by any means, electronic, mechanical, photocopying, recording or otherwise, without prior permission of the copyright holder.

Published by
5M Publishing Ltd,
Benchmark House,
8 Smithy Wood Drive,
Sheffield, S35 1QN, UK
Tel: +44 (0) 1234 81 81 80
www.5mpublishing.com

A catalogue record for this book is available from the British Library

ISBN 978-1-910456-16-3

Book layout by
Servis Filmsetting Ltd, Stockport, Cheshire
Printed by Replika Press Pvt. Ltd, India
Photos by Mick Rennison

'Keep on truckin' mama, truckin' my blues away,
Keep on truckin' mama, truckin' both night and day.'

Blind Boy Fuller

Contents

1

The Beginning: Miracle in Shrewsbury

'YOU'VE got no fucking chance!'

So spat my driving instructor as I sat waiting to take my HGV Class 1 test, on a cold wet day in Shrewsbury, 40 years ago.

It hadn't been an easy ten-day course. Me, long-haired and on the dole; my instructor, an ex-RAF drill sergeant, who didn't think it right that layabouts like me should get free training on a government-run scheme to help the unemployed. He never missed an opportunity to criticise me and my driving, while praising my fellow pupil to the point that he embarrassed him; his employers were paying hard cash to put him through.

Our breaks were spent listening to his tales of real men in the forces.

'Bring back National Service,' he declared. That would sort out wasters like me.

I have to admit, he did get to me. I can take criticism but I preferred it to be constructive. This was very personal. I was staying in digs, away from my home and loved ones, and more than once I vowed to quit and flee back home. But I knew that was just what he wanted, I also knew that I'd never get this opportunity again.

I'd qualified for the training course because, prior to my unemployment, I'd been a van driver for a couple of years. As I punched on up the M1 in a Trannie van, sleeping bag and cooker in the back, I longed for the chance to drive one of the big boys.

My father was a truck driver. Hand-balling 7,000 bricks on and off his wagon for Sussex and Dorking Brick Company. I went out with him as a kid, and loved every minute of it.

But when my chance came, it wasn't easy. It was like being back at school, with your least favourite teacher. At times he convinced me I really was useless, and punching far above my weight. They were dark days, followed by long, lonely evenings in my digs, revising the Highway Code.

I struggled to master the art of reversing. Never once during the entire course did I ever manage to manoeuvre the Atkinson Borderer and its 40' trailer into the coned-off box correctly. A task made even harder by my nemesis walking alongside the cab loudly yelling out 'Left hand down you bloody idiot! Now right! Are you really this stupid?'

Of course, my cab-mate did it perfectly with his first try on the first day of the course.

I sat in the waiting room waiting my turn. My stomach churned, the reversing manoeuvre playing in my head, over and over again. My instructor ignored me, no words of encouragement, no wishes of good luck.

When my cab-mate returned from his test victorious, he was greeted with handshakes and back-slapping. I left the room in silence with the examiner, shaking in my boots, sweaty hands clutching my provisional licence.

Alongside the Atki, the examiner explained the dreaded route of the manoeuvre. Drive forward, zigzag around a few cones, then reverse into the coned-off box. I felt as if I was in a trance.

Climbing up behind the huge steering wheel, I

wiped my hands down my jeans and fired her up. My thumping heart was drowned out by the throbbing Gardener engine.

I slipped her into gear, took the handbrake off and moved forward, gingerly passing the first cone. My mind was racing, my nerves were shot. Am I going too slow? Does that matter? Perhaps I should up the pace a little. In the mirror I watched as the flatbed trailer narrowly missed the cone, then I turned, gently, not too much, then swung back on the opposite lock. Round she went, then another turn, left a bit, then I was clear! Yes! So far so good. Now all I had to do was reverse on a slow curve into the dreaded box!

I repeat, I had never, ever, on the whole ten-day course, completed this manoeuvre successfully. I hadn't even got close.

She crunched into reverse and, riding the clutch, I slowly began to follow the imaginary line imprinted on my brain. Was that OK? Maybe take it off a bit more, no I needed to put it on, didn't I? I froze but the Atki crept on. Slowly but surely she slipped right into the box. Perfectly. Yes! Did you see that sergeant major? Did you bloody well see that? I couldn't believe it myself. Was I dreaming? No I wasn't, I'd done it! I'd bloody well done it!

The rest of my test passed in a haze. The emergency stop, then out onto the road. Keep checking the mirrors, look ahead, indicate, mind that bike, watch those traffic lights. In next to no time I was back at the centre answering questions on the Highway Code. No problem. Then the examiner was shaking my hand and congratulating me on passing my test!

With my pink pass slip in my hand, I floated back to the waiting room. My cab-mate shook my hand; he was genuinely pleased for me.

Sergeant major held out his hand. 'Well done,' he spluttered.

I ignored it, turning my back on him. I was buzzing, but my feelings for this man were so deep I didn't trust myself to speak to him. But then he said he wanted me to return to his office in the truck to pick up a certificate to say I'd completed the course.

'I've got what I wanted,' I sneered, waving the pass slip. 'Stick your bloody certificate!'

The classroom bully was lost for words. Now I didn't need him anymore, I saw him for what he was. An arsehole!

I felt so bloody good. Ten days of frustration vanished into thin air. I got my gear out of the Atki and caught a bus back to my digs, then a train home.

The certificate arrived through the post a few days later. It took pride of place on our toilet wall.

A life of travel took off for me in 1967, when at the age of 17 I joined the Merchant Navy as a catering boy. Sailing the world on passenger ships, oil tankers and tramping cargo boats; taking in such wondrous sights as the Great Wall of China, the Panama Canal and some amazing riverside Buddhist temples in Thailand; and visiting Australia, America and Africa along the way.

As a teenager you really don't appreciate the fact that you're being paid to travel the world. It's not until you're much older that it hits you just how fortunate you've been. At that age you think you know everything, you really do. You're the man!

Although the words 'naïve' and 'stupid' come to mind when I recall a time in Dacca, French West Africa, now known as Senegal. My shipmate and I, suited and booted with flash watches and rings, were easily tempted one night by a taxi driver offering to find us some girls. But we began to get a bit nervous when he pulled off the road and headed off into the jungle. Bouncing over the rutted track, things got worse when he pulled up in what looked like a small shantytown. He stopped alongside a large fire and we were immediately surrounded by a group of locals. As he spoke to them we locked the doors and wound up the windows. We were crapping ourselves!

'OK,' he said, pointing to some women. 'Jiggy jiggy!'

There was no way we were going to get out of that taxi, and we told him that in no uncertain terms. He just laughed. The women were beckoning us, one removed a breast from her dress and waved it at us. A couple of men began rocking the car. My, how the taxi driver laughed. We squirmed in fear.

'Back to the ship!' we demanded. He laughed even louder.

After a lot of arguing we struck a deal. We gave him all our money and our watches, and he took us back to the ship, laughing all the way.

Later, on the same trip, I had someone pull a gun on me in Bangkok. The blood that drained from my body, as we stood face to face in a bar room brawl, took several days to return. A few weeks later, in the China Sea, the ship went over on its side in a typhoon. She actually took water down the funnel and several portholes smashed, flooding cabins. On each of these occasions I really thought I was going to die. Very character building.

My first driving job, in the early 1970s, came about by accident. Living in my hometown of Horsham, in West Sussex, I'd left the Merchant Navy and was working as a salesman for Currys, the electrical appliances shop. The van driver who delivered the few

TVs and washing machines I sold had one accident too many and I took over his job. Temporary at first, but it was so much better being out and about all day, as opposed to being in the shop flogging fridges, that I leapt at the offer to become their full-time van driver.

In those days, firms often let drivers take their vans home. This meant I didn't have to have my own car, and Saturday nights usually meant a trip down to the seaside with all my mates in the back. This privilege came to an end, however, when we got stuck in the shingle on Bognor beach and needed a tow truck to pull us clear at 3am one Sunday!

I moved on to long-distance van driving when I went to work for Swedish Ericsson. Delivering telecommunication equipment all over the UK, I spent three or four nights out a week, usually in a sleeping bag in the back of a Transit van.

It was on one of these trips that I met my future wife Jo, in Barnsley. She was a single mum with two beautiful children, Justine, who was three, and 12-month-old Daniel. After a few months of long-distance romance we bought a caravan and moved it on to a farm in Llandegley, in Radnorshire, mid-Wales. Beautiful country and a wonderful lifestyle, but the scenery didn't pay the bills. Jobs were very hard to come by, especially for an English outsider.

After 12 months on the dole I was eligible for a government training scheme, and as an experienced van driver, I qualified for HGV training.

Soon after miraculously passing my test in Shrewsbury, we moved into a house in Knighton, Powys. It was there I got my first ever HGV driving job, with local tipper firm Brisbane's, pulling stone and ballast out of the local quarry at Nantmel.

As a novice (no experience at all), I started the first week travelling shotgun with an old hand called

Danny, in a ten-ton D-series Ford tipper. On the first day he proudly told me he'd been driving for more than 30 years and never had an accident. On the second day, while negotiating a narrow country lane, he managed to clip the grass verge with his offside front wheel and the truck slewed off the road and into a ditch. The Ford flipped over onto its side, the front wheel striking a concrete culvert and smashing back through the battery rack and into the diesel tank.

We came to a halt with me sitting on top of Danny, choking on the fumes that flooded into the cab. Battery acid and diesel is not a healthy mix!

I'd like to say I didn't panic, but I did – and so did Danny! I found myself standing on him and trying to open the passenger door. It was like a submarine hatch, I struggled to push it up and open, all the time listening to Danny screaming at me to get off him!

I eventually scrambled out and fell about ten feet to the ground. Danny soon followed and got his revenge by landing on top of me as I lay choking in the mud.

Danny elected to go off in search of a phone (no mobile phones back then) while I sat by the wreckage in a state of shock. He was gone for an hour or more, then it was a further hour before Brisbane's wrecker got out to us. Most haulage firms had their own wreckers in the 1970s. Even if you broke down hundreds of miles from home, they would be sent out to tow you in.

Crammed in the cab of the wrecker were three drivers with shovels. Once the Ford had been righted and dragged from the ditch, they toiled for a couple of hours to clear the truck's payload of stone chippings out of the ditch. Most of it went back into the tipper.

The trip back to the yard was almost as scary as the accident itself. There was no room in the wrecker, so Danny and I had to travel in the crumpled Ford. Although most of my nerves were already shattered, there were still enough left to scare the shit out of me as we swung about in the suspended truck on the hour-long journey back.

I stuck with the job and after a few months was eventually rewarded by being given my first articulated truck, commonly called an 'artic' by us truckers. She was a Leyland Lynx tractor unit coupled up to a 30' single-axle tipping trailer.

Of course I was a bit nervous, I'd not driven an artic since the day of my test. And more importantly, I'd never reversed one! But it was only a 30-footer for god's sake, and I knew most of the sites I'd be delivering to, so I was pretty confident I'd be able to blag it.

Stupid of me really, don't you think?

The worst day of my whole career, and possibly my life, began well enough. It was my first day out in the Lynx. Apart from the trailer cutting corners and the clutch being a bit on the heavy side, she was a nice ride. I loaded gravel at the quarry for a sewage plant, some 20 miles away. The boys at the quarry had been really helpful and drawn me a map, and from their description of the place, it seemed a doddle. A mile or so down a narrow country lane, but with a huge yard where I'd be able to swing her around to get out. Music to my ears! It was an unmanned treatment plant, just dump the gravel in the yard bin and push the delivery notes through the door.

Off I went full of the joys of spring, a proper truck driver at last! When I got to the lane and saw it was only a narrow dirt track, disappearing into a small copse, a familiar bead of sweat ran down my back. But the map marked it clearly enough; this was the place. Gingerly I left the road and went off into the countryside. After a few hundred yards, ditches appeared on both sides of the track. Deep, no-nonsense ditches, filled with water, run off from the plant. A tight right–hand bend had the trailer wheel crumbling the edge, clods of earth fell down the bank into the water. Then a left–hander. The boys at the quarry hadn't mentioned this. Another hundred yards or so and the trees were reclaiming the track. One branch bent my mirror around, another whacked the door. My anxiety levels were rising fast. Something wasn't right.

Another tight bend, then my heart lifted. Ahead of me I could see the treatment plant's iron railings through the shrubbery. I cursed myself for being of so little faith. This was the world of the tipper, country style. You went where you were needed.

The track widened a little, then ran alongside the high railings that surrounded the filter beds. It was when I realised that there was no opening in the fence that the enormity of the situation hit me. I was around the back of the plant! I couldn't get in and there was nowhere to swing her around!

Gripping the wheel I tried to contain the panic. I failed miserably. It spread from the tips of my toes to the top of my head in the split second it took me to realise that the only way out was backwards! More than a mile of snaking, twisting dirt track! And those ditches! I wouldn't get past the first bend!

I killed the engine and lit a cigarette. Climbing out I felt weak and light-headed, as if it wasn't really happening to me. A nightmare perhaps? Oh yes, please let it be some God awful nightmare! I pinched myself several times and it hurt. Shit!

The plant consisted of a couple of filter beds and a hut, set in a large compound, in the middle of some woods. What a stupid place to put a sewage works!

I struggled through the waist-high brambles and ferns around the perimeter fence until I found the entrance. It was alongside a neat tarmac road. A road that probably ran back up to the main road, and was probably the one I should have taken to get down here!

Walking back to the truck I verbally abused myself until I was hoarse. How could I have been so bloody stupid? Then I changed tack and blamed the boys at the quarry. Why didn't they say there was another turning? By the time I got back to the Lynx I was on the 'why me?' bit. What had I ever done to deserve this?

I had another ciggie while I studied the first bend. It would mean a blind-side reverse, vision obscured by branches, with only a foot or so to spare either side between tyre and ditch. And these 30-footers with a single axle, they came round so bloody quick. But the bottom line was, I couldn't reverse my way out of a paper bag! I'd have struggled reversing a Trannie van out of there!

After another ciggie in the cab, I'd calmed down enough to try and come to terms with my predicament. I needed help, I needed advice, but most of all I needed sympathy. This could happen to anyone, couldn't it?

For no good reason I pressed the horn. I let it rip for a good couple of minutes. Maybe I thought a fairy godmother would come by and magically turn the truck around. As you can see, my mind had gone.

I concluded that I needed to inform someone of my dilemma. Yes I'd probably get the sack, but that was better than facing up to the firm's other drivers once they'd heard. I could hear them laughing now.

But right now I needed help, and it wasn't going to come to me. There was no such thing as a mobile phone in those days and the chance of finding a working phone box within a ten-mile radius was very remote. I needed to find a house and borrow a phone.

In front of the truck, the track was now little more than an overgrown footpath. Covered with ferns and brambles, it was possibly just an old farm track. Hadn't seen a vehicle in years.

Could lead to a farmhouse, I thought, so I set off to find a phone. As I pushed aside the young saplings that had colonised the path, my mind raced. What would I say to the boss? More importantly what would he say to me?

After a hundred yards or so, I came into a clearing. A farm gate barred my way. On the other side of the gate was a field. A field! My heart leapt with joy! Even I could turn a truck around in a bloody field! Somebody up there was smiling on me at last! It had just been a fright to teach me a lesson, to test my reserve and character. And I'd come through it with flying colours, don't you think?

I promised myself I'd never ever get in such a pickle again. I walked back to the truck, checking out my intended route. It was a bit narrow and quite soggy and I'd need to flatten quite a few saplings. But it was all for a good cause, getting me out of the shit! I was on a bit of a high. Tomorrow I'd be laughing about this with all the boys back at the yard.

Firing up the Lynx, I edged my way forward. I'd only gone 50 yards or so when the wheels began to spin. I engaged crawler and switched on the diff lock. We progressed slowly. I say 'we' because all the time I was talking to her.

'Come on baby, you can do it. Don't let me down.'

I even apologised for getting her so muddy and promised her a good hose-down when we got back to the yard.

Saplings buckled under the bumper, branches attacked the wing mirrors and brambles clawed at the wheels, but we made it to the field. I opened the gate and slowly drove through.

Oh, did I tell you it was a ploughed field? No? Sorry, I was probably carried away by the euphoria of having found the solution. Or so I thought.

You know how it feels when you think a day can't possibly get any worse, and suddenly it's so bloody obvious it can? The familiar knot in the stomach, the cold sweat?

I kept her in crawler and charted a straight line for 50 yards or so. She ploughed on leaving some king size ruts in her wake. Then when I thought I had the room to swing her, I began to turn. I reckoned the larger turning cycle the better — after all, I did have ten tons of stone on.

As soon as the outfit broke out of a straight line, the wheels began to spin. The diff lock made no difference at all and we came to a halt. I rocked her backwards and forwards and eventually straightened her out. We moved off again. After a few yards I turned again, this time the opposite lock. Same result, but now she was really burying herself in soft clay. Again I rocked her out of it and made progress for another ten yards or so before trying another turn. This time she started to come round, trailer neatly following and I really thought we were on the home straight. Then the wheels began spinning again and we stopped.

I gave her some more revs, gently at first, then with some considerable force and anger. Then I lost it completely and by the time I gave up she was so deep in the mud the diesel tank was grounded and the spare wheel carrier had disappeared from view.

I cannot describe the despair I felt that day. It was a very long day, spent meeting a lot of very angry people. The irate farmer, who screamed loudly about damaged crops and land drains; my boss, who I thought was going to explode when he first arrived at the scene; and the driver of the first tow truck that arrived to pull me out. I say the first because his six-wheeler got halfway across the field before he too got well and truly stuck. I tried to tell him it was a bit soft, but he wouldn't listen. He called me names not appropriate for this tome.

Then there was the second breakdown truck. Came all the way up from south Wales, took him hours to get there, which for me were long hours spent in the company of some very agitated people. This wrecker was more like a mobile crane – four axles and a long reach jib. He sat on the edge of the field, hooked up to the first breakdown truck, who in turn was hooked up to me. Only it wasn't me anymore – the boss had brought along another driver. I just stood on the other side of the field watching, persona non grata. Stripped of all my pride and dignity.

In all it took the best part of six hours to recover the Lynx and get her back to the main road. By the time the boss dropped me off at home it was dark. His abuse rang in my ears for days and my self-esteem took many years to recover. By popular demand, I never returned to that job.

2

Down and Out in Radnorshire

I spent a while back on the dole. With a couple of kids to feed and electric central heating, we were often cold and hungry. I really needed to find work. Not just for the money, but for my salvation as well. The longer it took to get back behind the wheel, the harder it would be. Like falling off a horse, you needed to get straight back on again.

Then I heard on the grapevine that the local council wanted a tipper driver. At the interview I forgot to mention my previous escapade. I was so pleased to get the job that I also forgot to ask about the pay. Working a week in hand, it was a fortnight before I got my first wage. Shock horror! It was a take-home of just £33! I'd been getting £45 on the bloody dole! This represented a 25 per cent pay cut!

We'd struggled to survive when I was unemployed, no way could we manage with me working. After lengthy discussions with the dole office and Jo, I decided to quit. The dole office responded by reducing my money to £30 a week for six weeks, as a punishment for leaving the job without good reason!

We went from being poor to impoverishment in a matter of weeks. All because I took a bloody job! We lived on lentils and muesli. Buying them in bulk meant we always had something in the cupboard. Probably looked upon as a healthy diet nowadays but believe me, the exhaust gases they produced were far from healthy – although they did warm up the house a bit.

Some months later, a friend gave me a phone number. The guy's looking for a driver, he said. I scrounged a 10p off him and made the call. Yes! Couple of days a week, £1 an hour, cash in hand!

The guy picked me up from home the following morning in his car and took me to his yard. Now, on the phone, I'd only asked two questions. 'Is there a job?' and 'Can I have it?' Everything said after that was lost, as I was busy in my head listing what bills I could pay off with my first day's money. Might even stretch to a Sunday roast. What a treat!

So you can imagine the shock I got when we arrived at his yard. It was a knacker's yard! The first thing I noticed was the flies, swarms of them everywhere. The next was the beaten up old Bedford seven-ton box-wagon, sat in the yard, axle deep in a quagmire of mud and slush.

'Keys are in it,' said my new boss. 'Back her up on the bay.'

I waded over to the truck and clambered in. Firing her up, I swatted a few dozen flies against the windscreen with an old newspaper while she warmed up.

The loading bay was a waist-high platform at the rear of the slaughterhouse. Old, injured and diseased animals ended their days here.

After lowering the tailgate, I backed her on. As soon as I left the cab I came under attack from hundreds if not thousands of flies. Maybe they'd seen me swatting their mates in the cab and now, like something out of a Hitchcock movie, they were after revenge. I gritted my teeth and thought of the money. I had to be mercenary.

On the bay, at the sharp end of a conveyor belt, lay a pile of flesh about two metres high. Skulls, bones, and lots of grey intestines and thick yellow fat. I gulped my breakfast back down; I could so easily have turned and run.

'Skulls and bones on the front for Chester Zoo,' said the boss, ignoring the flies crawling over his face. 'The rest on the back for the dog food factory.' He handed me a pair of waders and a pitchfork. 'Hurry up.'

I'll spare you the gory details, of how the 'juices' splashed my face, and ran down the pitchfork handle and up my sleeves, or how many times I slipped and fell over in the foul smelling stuff. And of course the waders leaked, my feet squelching in bloody goo all day long. And the flies! Did I mention the flies?

The day was one of endurance, but I came through. I grabbed the cash but declined the offer of further work, citing a bad back. Back home I had to strip off at the back door before Jo allowed me in. My socks, shoes and shirt went straight in the bin. I soaked in the bath for over an hour before I felt clean again.

A few weeks later a driver agency, one of many I had signed on with, came up with a job. A hatchery, I'd be delivering day-old chicks to battery farms. All together now, 'ahh'!

When I arrived at the hatchery a D-series Ford box wagon sat on the bay being loaded. It was temperature controlled to keep the little critters warm. The wagon was already half full with roll-on cages stacked with large plastic trays.

I lifted a lid and peered in at my cargo. Ahh ... the fluffy yellow chicks, packed tightly together, looked up at me and cheeped.

A girl wearing a plastic hairnet and Wellington boots rolled on another cage and handed me the delivery notes.

'You're 200 short,' she said. 'You'll have to wait.'

'How long?' I asked.

'About ten minutes,' she replied. 'They're just coming out of the incubators now.'

While I waited I just couldn't resist taking another peep at my precious charges. The cheeping grew louder as I lifted a lid. They all looked up at me, eyes pleading. Did they think I was their mother?

They were so cute. Perhaps I could slip one in my pocket, sneak it home. They'd never miss it, there's thousands of them here. The kids would love a new pet. Fresh eggs too!

The girl returned with several large insulated boxes on a trolley. In them the chicks, dozens of them, were just hatching. Wriggling to break free from their shells. New life. They looked around at the world and shook themselves dry.

The girl inspected each one of them carefully before transferring them to the trays. Then suddenly she dropped a chick down onto the ground! It lay on its back, feet wriggling in the air, on the cold concrete floor. Then she dropped another, then another!

'What ...?'

'Rejects,' she said. Seeing the look on my face, she explained. 'If any are male, they're out,' she said. 'Like all males, they're good for nothing!'

I stepped back, I wasn't going to argue with her.

'And any with a speck or any colouring, like this one,' she said, 'the others would just peck at the speck until they kill it.'

It fell cheeping to the floor. Some had their shell stuck to them, this caused deformities, so they too joined the ever growing pile on the loading bay floor.

I couldn't stop looking at the wriggling pile of chirping chicks. Some lay on their backs, others had been injured in the fall, and some just sat there in silence.

By the time I got my quota there must have been maybe 30 or 40 rejects in a mound on the floor. None tried to run off, they just seemed resigned to their fate.

While closing up the truck doors I watched as the girl got a large rubber-edged broom and swept the sad looking heap of chicks off the loading bay and into a large skip. They joined hundreds more, cold and dead.

She told me they went to zoos and pet shops. Eagles and snakes just loved them.

At the battery farms I watched as the trays were emptied into enormous windowless sheds, never again to see the light of day. Maybe the chicks in the skip were the lucky ones. That day certainly gave me food for thought: the production line for our food is a lot longer than the supermarket queue.

I began to see my own hypocrisy. Everybody knows about the horrors of slaughterhouses and intense farming, and most of us think they're just awful. But we still carry on eating the stuff because ... well, that's just the way it is.

There was another transport firm in the town, Robert's, but I'd steered clear of them because they were long-distance boys, away all week. I felt I'd not got the confidence or experience to try that yet. Besides, Jo was expecting our third child and school runs (on foot) and shopping were now down to me. But times were bad and getting worse all the time. I felt I had to do something, so I started calling them. We had some very good friends and neighbours who could be relied upon to keep an eye on things with Jo.

Robert's of Knighton ran a huge fleet of trucks. The four-wheelers (two axles) were mostly Bedford TKs, running agriculture supplies to farms all over Wales and beyond. The artics were ERFs and Atkinsons, pulling 40' flat trailers on general haulage.

After weekly phone calls for a month or so, I struck lucky. At the interview I told them I'd worked for Brisbane's, but they didn't ask me why I'd left, so I never mentioned it. They did, however, ask me if I could rope and sheet. I felt that saying no would have been the end of the interview, so I said yes.

I know, I know, honesty is always the best policy, but I was desperate. My kids were starving. Well not really, but like I said, I was desperate.

So I got the job. Start off on rigids, then if I proved my worth I'd move up onto artics. Sounded great!

My first job was on an ERF four-wheeler (two axles), an A-series with a Cummins engine. It was already loaded with concrete roof slabs destined for a building site in Bromley, south-east London. More importantly, it was roped and sheeted. Great, I'd just look and see how the knots came undone and I'd be able to suss out how to tie them up. Yes? How difficult could it be? And the sheet looked simple enough.

I set off early on a wet Monday morning. Nervous of course, but fairly confident. The roads away from Knighton were winding and very narrow. Following B roads, I joined the A44 at Old Radnor and crossed the border into England shortly afterwards. After only one break for my brekkies, I made the site by early afternoon and tipped straight away. The site's labourers helped me take off the ropes and sheet, and although I tried to understand the intricacies of the knot, I just couldn't get my head around it.

The digger driver tried to show me by saying something about a rabbit that runs twice around a tree then disappears up his own hole. Eh? By now I was tipped and they wanted me out of the way, ready for the next truck. The ropes lay in a tangle on the floor, along with the hastily rolled up sheet.

I phoned in on the site phone and got my return load. Berkhamsted, a mixed agriculture load back to the yard. 'Get over there quick,' said the boss, 'and they'll load you today.'

I threw the ropes and sheets up on the bed and secured them down. Another driver at the site showed me the way on my map and I set off. I ran east around the South Circular, then onto the North Circular, picking up the A41 at Hendon.

The ERF was a lot quicker now she was empty. I felt under pressure to get loaded, I needed to show I was capable of doing the job. I arrived at the warehouse 20 minutes before they closed. The loaders were not that pleased to see me, they were ready to go home.

Two forklifts set about loading. Pallets of bailer twine, sheep dip and fertiliser, all different shapes and sizes. Then some plastic five-gallon drums of disinfectant were thrown on the back, along with some small farm machinery. In next to no time I was loaded and the boys were asking me to pull out onto the road, so they could lock up and go home. They put my heap of sheet and ropes on a pallet and stuck it on top of the load. I gingerly drove the truck out and parked in a lay-by alongside the warehouse.

Now what? Well the first thing to do was obviously get the sheet on. Climbing up on top of the load, I threw the ropes off and began spreading the sheet. Not too difficult, I just tugged and tugged, and pulled and pulled, working up quite a sweat. For a moment I thought the sheet wasn't long enough. Then I realised I'd spread it widthways, not lengthways. It was touching the road on both sides of the load. Damn!

This wasn't so easy to put right: the load was very uneven and not very stable. By the time I'd got it the right way around and covering the load, it was getting dark. I tucked the edge of the sheet under the pallets and drums. Now for the easy bit. I threw a rope over and hooked the looped end over a hook. Walking around to the other side I pulled the rope tight and . . .

To cut a very long story short, I tried every conceivable way to fasten down that load. Granny knots, sailor's knots and every bloody knot I'd learnt in the Boy Scouts. That bloody rabbit ran around so many trees, but it never once found its damn hole. I cursed and I sweated, I yelled at the truck and asked myself so many times why I'd taken the job in the first place. I did my best, then gave up.

By the time I climbed into my sleeping bag I was truly knackered. As I lay across the seats in the cab hotel — with newspaper and cardboard for curtains and an outside toilet — I wondered if truck driving was really the true vocation for me. I got very little sleep that night, my mind running through the problems I'd heaped upon myself.

At first light I was up and walking around the truck. What a bloody mess! A spider's web of ropes covered the sheeted load. They ran at all angles, from back to front and sideways. A hump in the middle of the load lifted the sheet clear on both sides, exposing the drums.

I tried tightening up the ropes by putting a stick through a knot and twisting it. It seemed to work so I used a few more in various places along the side.

I eventually set off and headed for home. I spent more time looking in the mirrors than I did looking out of the windscreen. Staying on the A41, I crept along going as slow as I could. Every roundabout and junction scared the crap out of me. Every few miles or so I'd stop in a lay-by and twist the sticks a bit more. At this rate it would take me days to get home.

I picked up the A44 at Oxford and tried to quicken the pace, only stopping at every other lay-by.

By late afternoon I was on the approach to Worcester. My clothes were soaked with sweat and I had a stiff neck from my constant mirror-checking.

Driving into the town, I noticed a bulge in the sheet, nearside. As my heartbeat increased, so did the bulge. When I stopped at traffic lights the bulge seemed to move.

Handbrake on, I leapt out of the cab. Running around to the side I saw the first of the drums drop off the truck to the ground. Then another, then another. I panicked, grabbing the drums and putting them on the pavement. They kept coming, like coins out of a fruit machine. By the time the load had settled, I had eight drums lined up on the pavement.

Being at traffic lights, I was causing all sorts of problems. The traffic jam stretched as far as I could see.

What now? Wait for the police to come along and nick me for an unsafe load?

'You got problems there, Drive!'

The understatement came from one of the crowd that had gathered to watch my misery.

'You can say that again!' I replied.

'Want a hand?'

I could have kissed him! He turned out to be a retired truck driver, and he should have had a pair of wings on his back. He suggested that I climb up on top of the load and pull back the sheet. Then he hand-balled the drums up to me one by one and I restacked them. He then loosened a few ropes, which he retied to secure the drums. I shook his hand, he wished me luck and then he was gone.

When I finally made it back to the yard, everybody came out from the office and the warehouse to revel in my misfortune. My, how they laughed!

Fortunately the boss also saw the funny side and I only got a gentle bollocking for 'misleading' him over my abilities. He then told some of the old hands to teach me a lesson. No, not a thrashing, they showed me how to rope and sheet. That bloody rabbit and his hole finally made some sense.

Baby Dylan arrived in the spring of 1976. Mother and son both brilliant. Being present at his arrival was the most amazing experience of my life. I enjoyed the time off, caring for our new addition, but needs must, and an extra mouth to feed soon had me back on the road. No such thing as maternity or paternity leave back then.

I stuck with the job at Robert's, and they with me, and was finally rewarded with an artic. An Atkinson Borderer with a 180 Gardener engine, just like the one I'd passed my test in. Going up the road in an 'Atki' made me feel like a real truck driver at last. Sleeping across the seats, on a door that I stowed behind the seats, was the real thing, and I had my own curtains that I jammed in the doors. Wagons with sleeper cabs were so rare that if you saw one at a transport café, it always had a load of drivers stood around it drooling!

I was still crap at reversing. When I arrived at a tip, I would park up outside and walk in to check it out before driving in.

One of Robert's' most feared back-loads was sheep dip out of Young's of Glasgow. A blind–side, dog's leg reverse onto the loading bay that gave me sleep-less nights just thinking about it. The sweaty hands, the palpitations, the sniggers from the one-shunt boys. Then you hand–balled one-gallon cans of dip, four high along the 40' bed, taking an hour or more, before roping and sheeting then trying to get a few hours down the road. Hard work and long days, but I loved it. Travelling all over the UK, there was a real sense of personal achievement when you retuned to the yard after three or four nights away.

By now I'd learnt to ask if I wasn't sure of anything. Older drivers on the firm were only too willing to offer advice and tips: where to park up for the night, and where not to; the best transport cafés; and how to get home on a dodgy night out. Most of it was very helpful but occasionally it verged on the suicidal.

The Atki was a great truck to drive. 'You're driving a legend,' the old boys told me. But she was a man's motor, if you know what I mean. The big steering wheel made your shoulders ache, and the accelerator pedal was so huge and heavy that to maintain a decent speed you had to physically hold your foot down by pressing down on your knee with your hand. Obviously this was very tiring so when the boys told me about the 'block', I was all ears. Apparently all the drivers had their own block. A simple pallet block, about six inches square, this was DIY cruise control in the 1970s. You would build the Atki up to 40mph – i.e., flat out – then you jammed the block down on the pedal and a convenient curve in the side held it down. Then all you had to do was steer. Brilliant! This was working-class innovation at its very best. Some of the boys were even thinking of patenting the design!

The first time I tried the block was on the M62, homeward bound from Hull with 18 tons of fertiliser on board. The M62 was a lot shorter then and only had two lanes, but it was a real roller-coaster of a road over the Pennines. The Atki would be slowly climbing up for 20 minutes or so, then she would run down again for ten, then back up again.

My limbs were going through the pain barrier when I suddenly remembered the block! They all said it was easy to fit, but I found it quite difficult, bending under the steering wheel, trying to jam it in, while at the same time keeping a straight line. But then, bliss! When my foot came off the pedal she kept her revs up and trundled on at 40mph. My aching limbs sighed with relief (is that possible?) as I settled back to watch the scenery.

Yes, it certainly pays to listen to them that's done it. I wasn't sure about some of their ideas though. Scotsman's fifth, for example, entailed throwing the gearbox into neutral at the top of a hill, and holding on for dear life all the way down. Some claimed to have had their motors off the clock doing this. Definitely not for me, but I was game for anything that made my life easier, and the block certainly made it a lot easier.

There was so much less traffic back in the 1970s. Sometimes it seemed you had the motorway all to yourself. The motorways were practically empty at nights. Night trunkers were a very rare breed and 'just in time' deliveries hadn't even been thought of.

About a mile ahead of me, a caravan appeared. No problem, at this speed it would take me ten minutes to catch up with him. I lit a ciggie, then stretched one leg, then the other. This is the life.

The road began to slope away and she slowly crept up to 50mph. This was fast, she began to shake a bit and I gripped the wheel tightly with both hands. I began to bounce in my seat. The Atki's seats were solid affairs with a bit of foam on top to cushion your arse. It was like sitting on a box. Perhaps that's why so many of us older truckers have bad backs. Along with being deaf in the left ear from sitting next to the engine all day.

At 55mph I decided to slip into the outside lane; the gap between me and the caravan was closing fast.

But checking the mirror before pulling out, I spotted a Guy Big J thundering up behind me. He was pulling a huge earth-mover on a low loader. The speed he was going probably meant he too had a block, or was in Scotsman fifth!

When he pulled out into the outside lane to pass me, I suddenly realised I was getting short on time and distance. I needed to slow a little, to allow him to pass me before I could come out around the caravan, which was by now only a few hundred yards away and closing. The Atki was now pushing 60mph and the whole cab was shaking.

No problem. I just coolly kicked the block off the accelerator and listened as the revs dropped down and she backed off to 50. I just needed to let the Guy by, then pull out into his slipstream and welly it passed the caravan. And who knows, I may well give the Big J a run for his money up the next hill. As he slowly edged past me the driver looked over and waved. I raised my hand, real cool like.

Even though I'd slowed dramatically, I was getting perilously close to the caravan, and the low-loader's rear end was still alongside my cab.

Time, I thought, for a bit of heavy braking. My foot hit the brake. Nothing! I pushed harder. Nothing at all, the pedal wouldn't budge! I kicked it, screaming abuse at it all the time. It was no good, the block had lodged under the brake pedal. Thanks boys. What do I do now?

The caravan was by now just feet away, and the low-loader was still alongside me. There was only one option: panic!

I swung on the wheel and yanked her onto the hard shoulder, avoiding a collision by a gnat's breath! As I undertook the caravan in a hail of dust and dirt, I looked down to see a woman in the car's passenger seat happily knitting away, completely unaware of the unfolding drama. Then she looked up in horror as I thundered past on the inside, just a few feet away.

With a little help from gravity up the next hill, then using the gears, I managed to bring her to a halt a mile or so down the hard shoulder. My heartbeat drowned out the throbbing Gardner. It was a good ten minutes before my knees stopped shaking enough for me to get out of the truck. I reached back in and pulled the block out from under the brake pedal, then I threw that bloody block out over the moors with such brute force and venom, I swear it flew over the horizon. It took a while before I felt fit enough to continue my journey that day.

After six months or so of earning a wage, we'd steadied the ship. We weren't well-off but our debts were manageable and the wolf was no longer at the door.

Jo and I just loved live music. As a teen, the first festival I went to was to see Bob Dylan on the Isle of Wight. I went to a lot of festivals back then, and most of them were free. Pink Floyd, The Who, The Moody Blues, and then Blind Faith in Hyde Park. Looking back, the 60s was a great time to be growing up.

So when we got the chance of seeing The Rolling Stones at the Empire Pool, Wembley, we jumped at it. Logistically, it wasn't easy. A babysitter, a borrowed car, a four-hour drive to London, see the show, and a four-hour hit straight back. But of course, it was well worth it. You need to treat yourself now and again. If not, what is the meaning of life?

It was while I was at Robert's that I finally saw sense and stopped eating dead animals. My mind had been troubled for some time now. The knacker's yard, the day-old chicks, the sheep and cows we could see grazing in the fields from our kitchen window. We ate very little meat anyway. Maybe chicken twice a week, and a hotpot now and again.

The final straw for me came in Newport, South Wales. I'd pulled into the cattle market late one afternoon. I was loading up out of the docks the following morning. Like most cattle markets in those days, it doubled up as a lorry park, once all the business of the day had finished. Most towns had one, and they always had toilets, and more often than not, a café as well. What more did we want back then?

The place was just winding down. A few trucks were pulling out, and one or two were still being loaded. I parked out the way, waiting for all the pens to be broken down and moved away. About 50 yards or so in front of me, a six-legger Foden was loading. Cattle were being driven up a steep ramp and into the wagon. They were very stressed and cried out, wide-eyed with fear. Any slow ones got a quick whack with a walking stick to hurry them on. When the wagon was packed full, there was still one cow on the ramp. There was no more room.

Four men in brown coats pushed and shoved at the hapless beast. One cracked it across the backbone with his walking stick. The cow yelled out, but try as it might, it just couldn't find any space to squeeze in. One guy then twisted its tail into a figure of eight, while two others, yelling and swearing, pushed with their shoulders on its haunches. The cow was going nowhere; there was simply nowhere for it to go.

Then suddenly one of the men lifted up the cow's tail and shoved his walking stick right up its arse! The poor animal screamed, a long drawn out guttural scream, as it leapt into the air, clambering up and over the animals in front of it. The sound was horrendous!

The men whooped with joy and quickly closed up the doors. There then followed a round of back-slapping, handshaking and the passing around of cigarettes. I was disgusted and sickened by what I'd seen.

And that was it. I've never eaten a dead animal since. I no longer wanted to be part of a system that showed animals so little respect. No one can say that they're doing it for me.

3

Tipping Disaster

IN the late 70s we moved back to civilisation. We all loved Wales, but with three growing kids, we needed to step up. We did a council house exchange with a family from Horley in Surrey. The new home had a large garden, a garage and a real fire. The planes from Gatwick, noisy and smelly, were quite a shock, but there seemed to be lots of job opportunities.

I needed to find work quickly. An advert in the local press took me to a firm in Godstone, on the A25. There was no M25 back then, the Surrey hills were still alive with the sound of nature.

The firm was a small three-truck outfit pulling sand and stone out of the quarries along the North Downs. At the interview the surly Bossman was very vague about wages and hours. I knew straight away the job was a bad 'un, but I needed a start.

In the yard I met a shifty mechanic called Dave. Our eyes never met, his gaze would be on the ground or flitting around nervously. He told me I was the seventh new driver in three months, the pay was crap and Bossman was an arrogant arsehole! He'd only been there four or five weeks himself. He showed me around my truck, a five-year-old Atkinson, with a Rolls-Royce engine and an Eaton Twin Splitter gearbox. It was hooked up to a 40' tandem (two axle) tipping trailer.

Paper cups, fag packets and chip papers littered the

cab, along with half a dozen logbooks. Few had been filled in properly, one driver had even signed his sheet 'John Wayne'!

Drivers' working hours are heavily regulated. Two loads of stone to Salisbury, then sand to Maidstone, in Kent. Then I loaded up for the next day, white sand out of the Buckland pit for Brentford in West London. I was absolutely knackered by the time I got home.

The first day was a very long one. You had to record your daily driving time along with your daily and weekly breaks. But enforcement was slack, the system was abused and, of course, the bosses would never stop you working all those extra hours for no extra pay.

I was back in by five the next morning, and got soaked just running from my car to the truck. It had pissed down all night. The local radio said there were trees down and flooding in Kent. The Atki's wipers struggled to cope with the blustering wind and rain as I set off for Brentford.

The tip was at a ready-mix plant, Day's of Brentford. I got another soaking booking in at the control hut on the weighbridge. The guy inside was a mountain of a man: fully bearded, Giant Haystacks the wrestler came to mind. He gave me my bin number and told me to keep it neat. No problem, I was a tipping pro. But come to think of it, I'd never tipped white sand before.

Driving around to the bin, I reversed up close. I was still crap at reversing, but there was plenty of room to manoeuvre. I got out into the rain and opened the tailgate. Water gushed out, followed by a dribble of liquid sand. Back in the cab, sat in my soggy strides, I engaged the PTO. The body of the trailer slowly began to rise up on its single ram. I watched in the mirror as the tailgate flapped. Water poured out before a sudden rush of sodden sand crashed down and the Atki shook.

When the body reached full height, I engaged the crawler and tickled the clutch out. As the springs lifted, I eased her forward some more and the sand flowed out. Keep it neat? Ha!

I pulled down the window and stuck my head out into the rain to admire my neat pile.

My heart leapt into my mouth as I saw the trailer wheels lift from the ground! In that instant, I knew we were going over. I grabbed the door handle to exit stage right. Too late, she went over with a thunderous bang.

With my right arm out the window and my left hand gripping the door handle, I managed to hold on as the forces of gravity tried to throw me across the cab. Log books flew at me and my lunch box hit me on the head. The windscreen shattered and came in, all in noisy slow motion. Then silence and stillness, the cab was crushed around me. I'd been here before, only this time there was no Danny swearing at me.

People appeared and pulled me out through the windscreen. The Atki was no more. The fibreglass cab had shattered, the hydraulic ram bent and the tipping gear had been ripped clean off. Although in a state of shock, I had escaped with barely a scratch.

The cause of the accident was plain to see. The overnight rain had saturated the fine white sand, and about a ton had stuck at the very top of the trailer. This had made her very unstable when the tipper body was at full height. As soon as I moved forward, she went over. I think I might have had a lucky escape there.

A cup of tea was thrust upon me and someone sug-gested I phone my boss. Couldn't someone else do it?

Bossman freaked right out. I was called things I'd

never even heard of. He reminded me I'd only started the previous day, now he'd be bankrupt! The abuse raged for the duration of the call. He told me he was on his way over. He also said he was going to kill me.

The Atki, laying crumpled on her side, looked like a slain dragon. Workers feverishly shovelled sand around the draining diesel tank, in an attempt to stop the flow reaching a nearby stream.

Work had come to a halt right across the plant. The trucks already in could not get out, and the queue of trucks outside had no chance of getting in. Most of the other drivers were sympathetic to my face as they all recalled their own tales of woe and destruction. Some saw me as a victim of weather and circumstances, others thought I was a tosser.

I crawled back into the wreckage to retrieve my gear. The passenger door had been smashed back into the engine hump. If I hadn't been holding onto the door I could have suffered some serious damage. I stuffed my wet sleeping bag with my wet maps and my wet clothes bag, then took it all into the canteen, away from the scene of the crime.

Big John, the guy from the weighbridge, was quite sympathetic. He said I could hide behind him if Bossman got violent. Waiting for him to arrive was like waiting for the hangman. When I saw his car pull into the car park, I gritted my teeth and went out to greet him. Shifty Dave was with him.

For the first five minutes or so, I hardly said a word. I just stood there, trying to weather the torrent of abuse that flowed from a very angry Bossman. He questioned my parentage, my ability to drive anything bigger than a bicycle and my very right to live. At one point I really thought he was going to lose it completely and hit me!

Once he'd got it off his chest, I followed, like a

shamed mongrel, as he walked around the fallen wagon. Shifty Dave suddenly got twitchy, nervous-like. Taking one look at the truck, he declared it a write-off and suggested we leave.

Bossman wasn't so sure. A crane would be needed to right her, then a wrecker to haul her back to the yard. He'd have to make some calls, lay it all on.

Shifty was still hassling to go. When Bossman snapped at him to shut up, he said he'd just go back to the car to get his toolbox, then he could remove the truck's batteries. Bossman gave him the car keys and Shifty scuttled off.

Bossman went to the plant's office and I took shelter back in the canteen. I sat with a cup of tea; wet, depressed and still in shock. I too wanted to get away from the scene of the crime, but I wasn't looking forward to the journey back to the yard with the boss. Things couldn't get any worse, could they?

Suddenly Big John came running in. He was waving his huge arms about and screaming loudly ... at me!

'Where's that little bastard? Eh? Where the fuck is he?'

He grabbed me by the scruff of the neck and demanded to know where my mate was!

'I ain't got no fucking mates!' I screamed in horror.

Two guys, carrying lumps of wood, had followed him in. Their eyes bulged and they were making excitable grunting noises. A table went flying as they surrounded me.

'Who? Where's who?' I yelled, cowering in my seat.

People crowded around to watch the kill. I'd survived a horrific accident, and now I was about to be murdered!

I thought they were after my boss, but no, they were after Shifty. I quickly explained he'd gone back to the car to fetch his tools. They ran out of the canteen, cursing and whooping. The crowd of rubberneckers followed.

I sat there, shaking in my soggy boots. What was going on?

Bossman came in. 'What the hell was all that about?' he demanded. 'Where's Dave?'

I explained how Dave was a very popular guy at the moment. I told him what had happened. He thought it best if we leave, he'd done what he had to do. I agreed, but before we got to the door, Big John and his buddies returned, crashing back in to the canteen.

Bossman's car had gone and so had Dave. They suggested we stay a while, they wanted some more info on Shifty Dave, or Bastard Bob, as they called him. Bossman was furious about his car. He demanded to know what the hell was going on. He'd only known Shifty for a month or so.

Big John calmed down a bit. He apologised for frightening me and we shook hands. We all sat down, and over a cup of tea Big John told us a tale of deceit and betrayal.

Ten years previously, a raid on a bonded warehouse in south London went wrong. Four guys were arrested and charged with armed robbery. To start with, the Old Bill had very little on them, but at the trial, the evidence against them was overwhelming and they were all found guilty.

Three of the robbers got eight years each. They were Big John and his baton-bearing buddies. The fourth got just 18 months. And guess who he was? None other than Shifty Dave, aka Bastard Bob! The disparity in the sentencing had led Big John and the boys to think that Shifty may well have done a deal with the police, and now, ten years down the line, by sheer chance, my accident had brought them all back together again. The reason Shifty was so nervous when he first arrived was that he'd clocked Big John and didn't fancy catching up on the old days.

The boys wanted his address. Bossman couldn't remember, but promised to phone them as soon as he got back to the office. Big John reassured him of the consequences if he didn't.

We finally left the site. Me like a wet and soggy tramp, sleeping bag over my shoulder, bag under my arm; Bossman, hands in the pockets of his nice warm coat, leading the way through the wind-driven rain.

The car was gone, so it was a bus, then the tube, then a train, followed by a taxi to the yard. Bossman said not a word on the entire three-hour journey; he totally ignored me. This suited me — what was there left to say? Besides, I'd slipped into 'poor me' mode. Somewhere during the day, I'd become a victim. I just wasn't quite sure when.

By the time we arrived back at his yard it was dark and still raining. My soggy skin was chapped and sore in so many places, it was a relief just to stand up.

Shifty had been and was long gone, collecting his own car and driving off with a boot-load of power tools and expensive kit! It transpired that the previous week he'd got a two-week advance on his wages, and he'd also taken money off the other two drivers, promising to get them cheap washing machines. Of course, his address proved to be as phoney as his name.

Bossman didn't actually sack me, and I never really quit. Nothing needed to be said. I just went home and never went back.

I often wonder if Big John ever caught up with Shifty Dave. Or if Bossman did indeed go bust. Either way, I don't really care. I must have learnt something that day, even if it was only to always sheet white sand!

I ran a while for Horley Road Services (HRS), just up the road from us at Salfords. Their fleet consisted of mainly Mandatas and D-series Fords, with a few Big Guys. It was a good steady job with lots of variety: steel, concrete and machines. I pulled my first trombone trailer, with 60' groins for sea defences, and my first low loader, albeit a small one.

One particular day I was driving down Morden High Street, in South London, when an elderly gentleman opened his car door while I was passing. It hooked into the trailer and crumpled back into his front wing. The windscreen shattered and the bonnet flew open.

I watched it all in the mirror, as if in slow motion, but there was nothing I could do. I wasn't going very fast at all and I'd stopped before the truck had past him. He was in a bit of a state, in shock, but unharmed. If he'd held onto the door, it could have been a bit different. His wife kept apologising to me, saying it was his fault for not looking. I knew that, but I really felt for the guy. He'd recently retired and had only bought the new Austin Princess a few weeks earlier. There was hardly a mark on my trailer.

HRS was a good firm to work for, although the wages never seemed to stretch far. These were the early 80s, the Thatcher years, and jobs were few and far between. Bosses could pay what they liked, and they did. Take it or leave it.

4

Three Loads Good, Two Loads Bad

I saw an ad in the local paper, 'Tipper drivers wanted. Good money for the right drivers.' I gave them a ring and spoke to the boss, Billy. He ran a fleet of bulk tippers out of South London. He had a contract with the local authority to carry domestic waste from transfer stations in the suburbs to landfill sites in Essex. The ease with which I was offered the job, over the phone, was frightening. But the potential earnings were nearly double of that at HRS.

The first day working for Billy started when he picked me up from home at 5am on a Monday morning. He took me to his yard in his shiny, cherry red Jaguar, with his own personalised number plate. He was a friendly chap, a cockney in his early 60s. He told me if I worked hard and kept my nose clean, I could earn some serious money.

The yard was actually a farmyard, the office a cowshed and the garage was a barn with a hole dug in the dirt floor for an inspection pit. But the trucks were real enough. Billy had an impressive fleet of a dozen or so Volvo F86s. Huge eight-legger (four axle) bulk tippers, maybe 14' high. All less than 18 months old.

With a modern styled cab, and plenty of power, the F86 was a driver's dream. It was in a class of its own. The dog's bollocks of the tipper world.

Billy gave me some keys and half a dozen log books. He pointed out my truck and told me to follow him to the transfer station, some ten miles away. No instruction, no advice and no training. Just 'Be quick!'

Inside the F86, I was panicking. I'd never seen so many gauges and dials. It was like bloody Concorde! And a couple of switches on the gearstick mystified me. It was a while before I even found the light switch.

Billy sat in his Jag, beeped his horn and shouted at me to get a move on. I stalled her twice before he came over and explained, briefly, how to use a range change. I'd never even heard of them. Sixteen gears? What? I'd never had more than six before.

We eventually set off, but it was a real struggle trying to keep up with Billy. He kept pulling over so I could catch up.

The transfer station was incredibly busy. Dustcarts from all over the borough ferried in the garbage, and Billy's wagons queued to carry it all away to the tip. The smell hits you first, then the noise and the dust.

Billy explained what was required of me. First rule of the day: don't weigh-in empty. Wastes too much time and clogs up the weighbridge, he said. The guy on the weighbridge knew the tare of all the vehicles. Just back up to the hopper and fill up to the brim. Weigh off, get your ticket and sheet up. Then run like hell to the tip, just the other side of the Dartford Tunnel at South Ockendon. Tip the load quick as you can, then run back here and do it all again. And again.

My required work rate was three loads a day, but Billy admitted that it wasn't easy. Being the nice guy he was, he'd give me a two-week trial. If after that time I was doing three a day regular, then the job was mine. If not, I was out on my ear.

For my first run I was going to follow another driver to the tip. When I introduced myself to him, he ignored my outstretched hand and just scowled as he told me that if I couldn't keep up, he wouldn't be waiting. Friendly fucker! Time was money, he said. Three loads a day and the money was very good, two loads and it was very crap!

Once loaded, I had to climb up a fixed ladder onto the top of the wagon. My feet sank into the rotting garbage as I struggled to spread the tangled sheet of netting over a stinking load. Not easy when you are so far off the ground. Then it was back in the cab and off we went.

With the M25 just a twinkle in some traffic planner's eye, the North and South Circular roads were the only way to circumnavigate London. The route to the tip ran along the South Circular, with several neat shortcuts through housing estates to avoid bottlenecks. The range change gearbox was giving me a headache. Buttons up or down? Gearstick in or out? I lost my helpful colleague at the first red lights. I stopped, he didn't! I carried on following the ring road. Ten minutes later, another of Billy's boys appeared behind me. I let him pass, then tried to follow. I managed to stay with him for a couple of miles before he jumped two red lights in a row and disappeared from view. Ten minutes later, I repeated the manoeuvre with the next of Billy's boys, and so on. They seemed to come along at regular intervals.

My problems really began when I finally got to the tip at South Ockendon. It was vast, acres of it, and so deep in rubbish, that when you drove over it, the truck rock'n'rolled like a ship in a storm. Walking round to open the back doors, my boots sank into the rotting garbage. The smell was horrendous. Screaming gulls zapped past my head as hundreds of them fought over the food scraps.

Now you may well remember, I certainly did,

that it was only a few months since that terrible day at the ready-mix plant. The painful memories were still there. My chapped thighs may have healed, but the psychological effects of going over are lifelong, believe me. You never think that something like that can happen to you. But once it has, your confidence is shot. If it's happened once, you know damn well it can happen again!

As I raised the tipping body, she leant precariously over to the offside. My bottle went, I lowered the body and moved to a different spot. The same thing happened, only this time she leant the other way. I watched other trucks tipping. They leant just the same, but their drivers didn't seem to worry, they just carried on tipping. Then shot off for their next load.

I eventually gritted my teeth and went for it. Holding tightly onto the steering wheel with sweaty palms, I was convinced she was going to go over. It was nerve-racking, but I made it. Pulling gently forward, she disgorged the waste and I drove back to solid ground.

That first day I only did one load, then returned to the transfer station and loaded for the next day. I took the truck home with me, parking it in the Chequer's Pub car park, alongside another of Billy's wagons. I was absolutely exhausted and I stank of rotting garbage!

The next day I left home at 4am so I could be at the tip when it opened at 7am. But I still struggled to get in two loads and a reload. Most drivers seemed to be doing their quota – a few even claimed to do four a day!

On my third day I got a puncture at the traffic lights on Bromley High Street. I had never changed a wheel before and the task took me more than two hours. During that time, at least ten of Billy's trucks passed

41

by. Not one of the drivers stopped to see if I needed any help. Time was money and everyone was chasing the clock. The mantra was: 'Three loads good. Two loads bad.'

I'd been running for Billy for a week and a half before I did my first three-load day. I was by now quite fluent with the range change, and my tipping times improved as my confidence grew. I ran non-stop from four in the morning to nine at night. I carried a flask and some sandwiches and ate on the move. Need the loo? Go behind the truck on the tip.

Billy made me permanent and I began to earn some serious money. Cash in hand, of course! But I was absolutely knackered and had a constant headache. It was go, go, go as you pushed your luck every minute of the day. Red lights? What red lights? They were amber, honest officer! I carried three log books. Two ran together during the day, the other was the master copy, filled out at the end of the day.

All of Billy's wagons had battle damage, but he never went on too much about it. Unless of course the motor had to come off the road, then there was hell to pay!

One of the delights on the run to the tip was the Dartford Tunnel. Then it was just a single bore, with two-way traffic. As you thundered on through, the chances of meeting one of your stable mates coming the other way was pretty high. With wing mirrors just inches apart, you never waved, or even nodded to each other. You just fixed your eyes solidly on the road, gripped the wheel with both hands, and tried to hold her straight as the down draft from both wagons collided alongside you. All the wagons carried spare mirror glass.

One sunny day, after emerging from the tunnel southbound, I noticed a couple of kids on the hard shoulder, under a footbridge. They must have been about 11 or 12 years old, and wore school uniforms. Thinking that they were trying to cross the busy dual carriageway, I backed off and moved out a lane. One of them had a plastic carrier bag, and when he began swinging it around his head, I knew just what was coming, but could do very little about it. As I drew level with them he threw the bag, weighted with half a brick, at my windscreen. The fact that the Volvo was one of the few trucks around with a laminated screen, probably saved me from serious injury. The screen shattered but stayed in one piece, and I was able to pull her onto the hard shoulder. By the time my trembling legs allowed me to get out of the cab, the kids were long gone. The little bastards could have killed me! But, to be honest, I was more pissed off about losing a load. I pushed out the screen and slowly took her back to the yard.

In the three months I worked for Billy, I never had less than two punctures a week. One week it was four, including two in one day! Running over the tip you would get spiked by bottles and lumps of metal. Once I came off with a pair of scissors impaled in the side wall of a front tyre. I got my wheel change time down to 20 minutes, others claimed to do it in ten. A puncture could easily mean a lost load, then I'd have to work Saturday morning to make up my quota.

A lot of the drivers were ex–army, so they were used to hard work and doing as they were told. Others, like me, were just passing through – earn some hard cash, and get out before it kills you! None of them were particularly friendly, some were even downright hostile. I learnt pretty early on not to ask too many questions or, better still, don't ask any at all.

One thing that puzzled me from day one was why the weighbridge operator, a council employee, would often wind on the clock by hand, giving me a ticket showing an extra couple of tons. This all became clear one morning when I saw Billy slip him a brown envelope. We drivers got paid by the load, Billy got paid by the ton. There was no weighbridge at the tip, you just got your ticket signed.

The mystery of the guys doing four a day became clear when I realised they weren't tipping in the same place as the rest of us. They were running to another transfer station, much closer than South Ockendon, and giving the council foreman a few quid to tip there instead. The extra load a day almost doubled their wages, no questions asked.

Another opportunity for Billy came when a dustman's strike hit South London. Rubbish piled up high in the streets, and when, after a couple of weeks, the strike ended, Billy got the contract to shift it.

The rotting garbage was loaded onto our trucks by JCBs. In the absence of a weighbridge, the JCB driver, a council employee, would issue a handwritten ticket for one full load. We then ran up to the tip. Billy was on-site a lot of the time and was very friendly with the JCB drivers. This might be why some trucks, the chosen few, pulled away with only half a load, enabling them to call in at the transfer station on the way to the tip and top up their loads, thus getting a second ticket. Billy, and the driver would then get paid two loads for the delivery of one! Only chosen drivers were invited to participate in Billy's many scams. And, with hindsight, I'm pretty glad I wasn't one of them.

I quit the job after three months, on health grounds. I was bloody exhausted! My lifecycle was work, work and more work. With a little sleep in between. Some days I honestly didn't know what day it was. I would go into a trance, with just one aim: one more load!

And it didn't bother me at all to drive past my puncture-stricken buddies, even if they were struggling new boys! They'd learn. And when a new driver asked to follow me to the tip, I heard myself saying 'Well OK, But if you can't keep up ...'

My family life had all but disappeared and I permanently stank of garbage. I got used to it, but Jo and the kids never did.

But yes, I did earn some bloody good money. In those three months we got a motor, an old Bedford minibus, and a new washing machine and TV. And we all had a great Christmas, Santa did us proud. But these were all material things, my health and sanity had to come first.

A month or so after I'd stopped running for Billy, the law came knocking on my door. Billy had been busted! Unbeknown to me, and obviously Billy as well, the Old Bill had been tailing the trucks for six weeks. I was in the clear because I'd never been involved in the dodgy runs or scams. I knew nothing, and I told them as much. But Billy, five of his drivers and three council employees all got time. Billy got five years, the rest between four years and 18 months a piece. Others involved got big fines and probation.

Billy had ripped off the local authority for more than £100,000 over a two-year period. A lot of money now, but megabucks back then. His other misdemeanours included stolen tyres, red diesel, dodgy insurance and tax avoidance.

If there is to be a moral to this story it must be something along the lines of 'thieving greedy bastards always get caught'. But for me, the lesson must be that a healthy mind and body take precedence over a healthy bank balance any day.

5

On the Agency

AFTER a few days' rest and recuperation, I signed on for an agency in Croydon, Plan Personnel. They set the pattern for most of the agencies, dozens of them, that I would work for throughout my career.

'Class 1? Great! Come in. Coffee? Sit down, have a chat.'

They wanted me to be their friend, not an employee. They loved me, they thought I was great.

'Work? Yeah, we've got loads. Start Monday? No bother.'

After a couple of weeks of broken promises, lots of phone calls and a few visits to their office, I finally got a bite.

'Could I do long-distance?' he asked. He sounded desperate, I got the impression someone had let him down. 'Two drops, Dublin and Belfast. Tomorrow, out of Purfleet. And don't forget your passport.' Wow, my first trip abroad!

Jo was a bit concerned. The evening news had been full of 'the Troubles' in Northern Ireland, for weeks. Bomb blasts and knee-capping seemed daily events. Perhaps I only got the job because no one else wanted to go.

The evening was spent pouring over maps, and checking out my route. I was so excited I hardly slept.

The kids all got up to wave me off. My trusty Bedford minibus had me in Purfleet by 6am where a ten-ton Bedford TK was loaded and ready to go. I was given the delivery notes and the ferry booking times, Holyhead to Dublin out, then Larne to Stranraer back. Then I was given the keys and told to hurry up. If I missed the Holyhead ferry it was an eight-hour wait for the next one.

Now if the Volvo F86 was at the top of many a driver's wish list, the Bedford TK wouldn't even make that list. Compared to the Volvo, the Bedford was a museum piece. The cab was small and sparse, my sleeping bag and kit seemed to fill most of it. The roof was so low you were forced to lean over the steering wheel. And I'd gone from a 16-speed gearbox to a five-speed one.

Setting off, I struggled around the North Circular Road towards the M1. It was hard to make decent progress, the paperwork said eight tons, but I felt she was pulling 20. She was a real bumpy ride, potholes sent judders right through her, and me. My poor back! The power steering in the F86 had spoilt me, my arms were aching after the first few roundabouts.

The motorway brought some relief, although with a top speed of 50mph I never passed a thing. After a welcome cuppa at the Blue Boar, Watford Gap, I pushed on up the M6. Just before Birmingham, I got pulled over onto the hard shoulder by the police. I thanked my lucky stars I'd filled in the logbook before I set off.

After checking my logbook, they wanted to know why I was going so slow. What was I carrying? What were the weights?

'Only eight tons,' I protested, handing over the paperwork.

They weren't convinced and suggested I followed them to a weighbridge.

They pulled away and I turned the key to follow. Nothing! The key just clicked. I tried again. And again. Nothing. I sat there for a while, then suddenly the police car was reversing furiously back up the hard shoulder towards me, blue lights flashing.

An irate policeman jumped out. 'What the fuck you playing at?' he spat, pulling open my passenger door.

I tried to explain what had happened, turning the key to demonstrate. She fired up! Bitch!

'Yeah?' said Mr Plod. 'Just follow us, no more monkey business, OK?'

He slammed my door shut, then off we went.

At the public weighbridge, a steel works just off from Spaghetti Junction, it proved I was indeed only carrying eight tons, the paperwork was correct. The disappointed cops then pointed out that it was actually the leaf springs that were the problem. They were flat! Not a curve to be seen. Well and truly knackered!

The boys in blue had a chinwag in their car. To put me off the road, they'd need to wait for a Ministry-appointed mechanic to come out and confirm their diagnosis. And they were getting awfully close to the end of their shift. Justice prevailed: they let me go with a warning.

I set off again with a smile on my face, and the knowledge I was never going to make that ferry. But at least now I knew I had an extra eight hours to get there, I could stop stressing and enjoy the ride.

Following the A5 through Shrewsbury, my mind slipped back to the Sergeant Major. What a bastard! And that bloody reversing manoeuvre! Stressful times. A lot had happened to me in my career since then, but I was still crap at reversing!

The A5 runs right through Snowdonia. It was just

amazing, the views, the soaring buzzards; it made me question why we'd ever left Wales.

Crossing the Menai Bridge onto Anglesey, I arrived at Holyhead with only three hours left to wait for my ferry. I phoned in and told them about the police delay, and the state of the wagon. They weren't very happy, but then neither was I.

I had time for a leisurely meal at the ferry port before boarding.

Now, this may surprise you, me being an old sea dog and all that, but I do get seasick. It's not really a problem, I just feel better if I'm horizontal for most of the trip. When I went to sea, I always felt crap for the first two or three days of the trip. You couldn't lay down then, you just worked through it. But on this short crossing, a little over three hours, I just had to grin and bear it.

On arrival I grabbed a few hours' kip in the docks before setting off. I soon found the tip in Dublin, just outside the port. Half the load, (groupage), was duly hand-balled off by me onto pallets. A quick cuppa and a sandwich, then off I set for Belfast. The coast road up through Drogheda and all the way to Dundalk had fantastic views out over the Irish Sea, but was the most appalling road I'd driven on. Pot holes everywhere, big deep wheel-bouncers that oncoming cars just swerved around, forcing me to take avoiding action – scary! Most vehicles on the road seemed battle-damaged. Some cars had no lights, just empty sockets, others had broken windscreens and damaged bodywork hanging off. I kid you not. Didn't they have MOTs over here?

A tipper struggled to overtake me on the approach to a bend. As he drew level with me I noticed he didn't have a nearside wing mirror! How's he going to know when he's passed me? I wondered, seconds before he swung across the front of me, nearly putting me off the road!

Approaching the border, the Irish Tricolour was everywhere. Every house and building seemed to have their own flagpole.

Across the border in Newry, it was the Union Flag. Hanging from chimney pots and bedroom windows. Telephone boxes and kerb stones were painted the red, white and blue of the Loyalists. If you didn't know better, you'd think it was carnival time. Two religions, two cultures and hundreds of years of history kept these communities apart, along with a lot of bigotry.

Although a devout non-believer, I respect all faiths and beliefs. But what I find hard to comprehend is the willingness of humans to kill other humans in the name of their merciful gods. How can killing an innocent child advance your cause? Violence perpetuates violence. Bombs in supermarkets and train stations turn freedom fighters into terrorists.

My tip was on the northern outskirts of Belfast. I passed through several checkpoints, manned by young soldiers and armed police. They looked tense and very serious. No smiles or nods, just a gun and a glare. The next vehicle in the queue could be the one that kills them.

IRA and Loyalist murals decorated whole ends of houses, sometimes on opposite corners of the same street. Guns and gravestones seemed common themes.

My offloading this time was assisted by a young apprentice. He seemed a real nice guy. We discussed politics and he told me that the day would come when all of Ireland would be reunited. He had a brother in prison and had lost an uncle in 'the Troubles'. His argument for the cause seemed justified and heartfelt.

'But what about the people here who want to stay in the Union?' I asked.

'If they want it that bad,' he replied, 'why don't they just move to England?'

I ran north out of Belfast and caught the night ferry from Larne. A stormy three-hour crossing to Stranraer left me exhausted, so I took a kip for a couple of hours before leaving the port.

It took me nearly 12 hours to cover the 430 miles back to Purfleet — although this did include a kip for an hour or so back at the Blue Boar.

On arrival back at the yard they just took the keys and the paperwork off me and I went home. For some reason I half expected a 'well done', or maybe a 'thanks'. But no, apart from a dig at me for taking so long, they said very little. As an agency worker, I was going to have to get used to being a number. A very small cog in a very big wheel. To operate smoothly, you had to know your place.

And it wasn't just the bosses who treated agency drivers with contempt. My next job from Plan Personnel was for Sainsbury's, at Hoddlestone, near Croydon.

Now I've been in the union since I first left school and started work. We workers need the union's organisation and clout to protect our rights and establish new ones. United we stand. So we're all brothers together, right? We're all fighting for the same things, right?

On my first day on the job I was approached by the union shop steward. No hello, no handshake, no introduction.

'Agency?'

'Yes, I'm Mick ...'

'In the union?'

'Yes, I ...'

'Card?'

He studied my union membership card for a moment, then returned it and walked away. I felt angry at his aggressive stance. Why the bad vibes?

I worked out of that depot for a fortnight and, believe it or not, that same shop steward asked to see my card every single day. Why? Because he could.

The union was strongly opposed to the use of agency drivers. They wanted real jobs to be created. Great, I can go along with that. But why create such oppressive conditions for fellow drivers? We're all supposed to be on the same side, aren't we? Several drivers treated me as if I threatened their very livelihood. Any vehicle damage was always 'those bloody agency drivers'.

The agency itself was very fickle. If they had lots of work, they loved you. If work dropped off, they'd lie to you to stop you looking elsewhere. It was always 'I'll call you tomorrow with the details'.

But that's the world of the agency driver, that's how it was and still is, to some extent.

In the late 70s, health and safety was seen as a joke by most firms. Any visits by officials was always known well in advance and the appropriate lip service paid. Steel toe-caps were optional, hi-vis unheard of, and if you hurt yourself it was your own bloody fault! There was no system of training, anybody from the cleaner to the tea boy could jump on a forklift and unload the wagons. The only risk assessment taken was whether you'd finish on time to go home.

I started a week's holiday relief at a builders' merchants in Redhill. At 6am every day, a loaded wagon would be waiting for me with half a dozen local drops. Every day a different motor. Bricks, cement, timber and plumbing gear. All hand-ball and usually delivered to small building sites and private homes.

The first couple of days went well enough, then I turned up one morning to find a brand new Volvo six-legger (three axle rigid, i.e., six 'legs') waiting for me. And it had a hi-ab, an extending hydraulic jib arm with a hook and chain, something not often seen back then. The load consisted of the usual hand-ball and a couple of one-ton bags of sand.

The controls for the hi-ab, just behind the cab, looked complicated and confusing. I asked the yard manager what lever did what? He didn't know either. He played with it for a while, sending the jib this way and that. Up and down, sideways, then in and out. Then he slid out the stabilising legs, then slid them back in again.

'Simple,' he said. 'Now get a move on.'

The first couple of drops went OK – 200 bricks to a new shop fit-out, then some timber to a garden centre. The sand, in big one-ton bags, was for a house extension on a housing estate. It wasn't an easy tip. The wagon was half on the pavement, but still blocking the road. A tree in the front garden meant I couldn't get close, and it started to piss down with rain.

Legs out, jib hooked onto the bag, and slowly lift. As the bag came off the bed of the wagon, all these bells started ringing. What's all that about, I wondered? I lowered the bag, then tried again. Same result. I read the plate on the controls and it said the bells would ring prior to the wagon falling over. Something to do with the forces of gravity. Oh!

Using the jib, I dragged the bag up the bed, closer to the headboard, and tried again. Yes! No alarm bells. Well, not until I'd swung out clear of the motor and over the garden fence. So I quickly lowered the bag down. A bit too quickly. The truck rocked wildly and those bloody bells started off again. But the bag was down and as the rain ran down the back of my neck, I went for the next.

By now several cars had backed up down the road and a few beeping horns showed they were getting impatient.

I went for the second bag like a pro. Hook up, swing out, ignore the bells, over the fence and down. Yes!

I scurried around, unhooking the bag, and swinging the jib back over. Legs wound back in, paperwork signed, then back in the cab and out of the rain. Job well done!

My next, and final drop, was at the back of Gatwick Airport, just a few miles away. In my head, I was already looking forward to my dinner and the footie on the telly.

A dual carriageway ran past Gatwick, alongside the runway and terminal buildings. Halfway down, a covered walkway spans the road, taking passengers from the car parks into the terminal. As I passed underneath, an almighty bang was followed by crashing sounds as the truck veered violently across the road. I came to a halt, my poor heart thumping on my rib cage.

I think it was when I noticed the hydraulic oil, pouring down the windscreen, that I realised I hadn't lowered the jib properly. It had struck the walkway and ruptured the pipes! Oh shit!

Getting out on shaky legs, I saw a couple of chunks of concrete in the road. The jib had smashed several large pieces of reinforced concrete off the walkway.

My panic-stricken mind took over. Thankfully there was no other traffic about. My first thought was to clear off before the law turned up. I cleared the fallen debris from the road, then drove the truck clear, pulling into a lay-by, some hundred yards down the road.

The damage to the hi-ab was significant. The pipes were all split and the bed of the truck was covered in oil, as was the whole of the cab. And worst of all, I couldn't lower the jib. It was stuck upright, maybe 17' or 18' in the air.

I had a ciggie and tried to come to terms with my predicament. How could I have been so stupid? This is what happens when you don't know what you're doing. And when you rush. The waiting cars, the pouring rain, the next drop. But the bottom line was, I'd cocked up big time.

Now those of you that know Redhill will also know that it is surrounded by low bridges. A lot lower than I now was. With no cab phones and the embarrassment factor cutting in, it never occurred to me to call for help. It was my responsibility to get her home.

It was a nightmare. With no *A-to-Z*, or indeed any map at all, it was a case of trying a route, then backing off and turning around at the first sight of a bridge or low telephone cables. And there are a lot of them surrounding Redhill and my depot. A hell of a lot!

My disaster occurred around 4pm, I finally made it back to the yard by 7pm. To find my angry manager waiting for me. He got even angrier when he saw the damage and heard my tale of woe. As you will have probably realised, I never went back to that job.

Looking back it's easy to say I shouldn't have been allowed to operate the hi-ab without any proper training. But at the time, that's the way things happened, and I had to accept it was all my fault.

6

United We Stand

IN 1980 we upped sticks and moved to the seaside. East Preston, near Rustington, West Sussex.
A couple of hundred yards from a beautiful beach,
with the South Downs as a backdrop. And no aircraft
noise or smelly pollution.

One of the conditions of our council house swap
was that I had a job to go to. A few weeks earlier I had
an interview at the local council depot and was taken
on as a relief driver. Now you may well remember that
whenever I get a new job, I always seem to forget to ask
relevant questions. So it wasn't really a surprise, when
on the first day, I discovered I was giving relief to the
driver of a dustcart. But while he was driving, I was
running behind with a bin on my back! Yes, I was a
dustman. In those days, the bins were not put out. You
went into the garden, picked up the bin, slung it up onto
your shoulder, back out to the cart, tipped it in manually then retuned the bin to the garden. If that wasn't
hard enough, it was job and finish for the bin-men. So
the cart would just roll on down the road slowly and
the guys just ran after it. They'd usually finish by about
midday as most of them had second jobs as taxi drivers.
That first day nearly killed me. It was mid-summer, a
long hot day, and I'm not built to chase dustcarts with a
bin on my back. Then, while the driver and the rest of
the crew took a break in the cab, I drove to the tip and
unloaded – so I never got a break at all.

At the end of the day my back was killing me, and to the disgust of my loving family, I returned home with that familiar smell of rotting garbage! I lasted the week before quitting on grounds of exhaustion. Defeated and broken, I vowed to ask a few more questions before I took on my next job.

Which I did, and a Saturday morning interview landed me with a job the following Monday.

J.S. Salbstein's was a fruit and veg wholesaler based in Worthing. An old family firm, the office walls were lined with black-and-white photos of horse-drawn carts pulling potatoes and apples to market. They had several farms and orchards in Kent. Daniel Salbstein was the third generation of his family to run the firm.

About a dozen trucks supplied veggie shops and supermarkets from Bognor Regis to Brighton. Then, after the deliveries, you would have collections from local farms and nurseries.

My first wagon was an old Dennis, but most of the other trucks were ten-ton Leylands. A six-legger Volvo and an eight-legger Mammoth Major ran in potatoes from Suffolk and fruit from Covent Garden. The trucks were all immaculately kept, with old-style paint jobs and logos projecting the image of the firm.

A 6am start would see all hands loading the trucks. A caller stood on a box and yelled out the orders, and we drivers would collect the goods from around the warehouse on our trolleys and pass them up to the driver, on the back of his truck. Once loaded, the trucks pulled out. This meant that if your truck was loaded last, you helped load ten or more wagons before you set off.

Once you got into it, it was a pretty good job. You weren't pushed too hard, it was a nice steady number. Daniel wasn't a bad boss, although at times he could be a bit aloof.

At Christmas they had a work's dinner in a restaurant, and all the employees got a potato sack full of fruit and veg along with a turkey. I passed mine on to my brother. In the summer there was a kid's day out to Butlins, just down the coast at Bognor Regis.

I stayed with the firm a couple of years and progressed to driving the Mammoth Major. A real beast to drive and now, I'm told, a legend. She was a short wheel base and apparently there were only three or four of them in the country, two being with travelling fairgrounds. At the time it was just another truck, but one that took me off shop deliveries. Now I was collecting spuds from Suffolk, apples from Kent and early morning collections from Covent Garden. It meant the occasional night away, sleeping across the seats.

Spuds were hand-balled off the fields. You stood on the back as a conveyor belt delivered them up to you, one by one. You grabbed a bag, turned and dropped it on a pallet. Then turned back for the next. If you dropped one, or it landed wrongly, you had little time to correct it as the next bag was on the way. The guys on the field could get pretty pissed off if they had to keep turning off the belt while you sorted it out. It was OK for them, once I was loaded, they went home. I had to rope and sheet, then drive down to Brighton Market and offload them.

Being a short wheel base, although she could haul 18 tons, you could only get 16 pallets on the bed. So with potatoes you had to hand-ball a couple of tons across the top of the load. And because the spuds overhung the pallets, you'd have at least half a pallet hanging over the back end of the wagon. A couple of long pieces of angle iron under the rear pallets stopped them snapping, and lots of ropes and sheeting kept everything in place.

In the early 80s, the Tachograph was introduced for all trucks over 3.5 tons. Up until then we had log books that were supposed to be filled in at regular intervals in our working day. But of course it was much more convenient to do it at the end of the day, and if you wanted to run bent then it was easy. Drivers would often run two log books simultaneously, and there was little chance of getting nicked.

Called the 'spy in the cab' by many irate drivers, the 'tacho' sat behind the speedometer on the dashboard. You fitted a paper disc called a 'card' at the start of your working day and it automatically recorded your every move, including your speed, distance and how long you stopped for that kip up the road. Drivers hated it and soon found ways around it. You could always remove the fuse or simply change cards, but the chances of getting caught increased dramatically. Thankfully at Salbstein's there was never a need to run dodgy. You were never pushed to do anything illegal.

The monthly union meetings were casual affairs. Poorly attended, they often just turned into whinging sessions with very little positive input or outcome.

Originally the shop steward was the foreman. A real nice guy, Ron was in his 60s and had been with the firm all his working life. But when there were any serious rumblings of discontent, he found himself torn between workmates and master.

One of the meetings got quite heated and Ron had had enough, so he resigned and somehow I was elected shop steward. I never planned it and it really was a lot of hassle as everybody and his dog felt they could now whinge at me.

The main bone of contention was washing the trucks. As I said previously, the trucks were immaculate, well-maintained and spotless. Drivers washed their trucks once a week. No problem there, that's an expected part of the job. Us drivers like a nice clean truck.

But once a month the trucks were serviced and here lay the gripe. The trucks would be jacked up onto blocks in the yard and the wheels removed. Then they would be steam-cleaned by the driver, who was also expected to climb underneath and jet-wash the engine and the gearbox. Nowadays this would be a big no-no, on the grounds of health and safety. But then you did it or you quit.

The situation wasn't helped by the guy who ran the garage. The fact that he was ex-RAF (yes the scars were still there) – and an arrogant SOB to boot – meant we clashed almost daily. Once he told me I didn't need lights on in the rain. His previous employment was with a bus firm, say no more. One driver was sacked after an altercation with him: he had him pinned to the wall with a hand around his throat before we managed to drag him off.

The drivers I now represented made it pretty clear to me that the monthly service was not a job for drivers. The garage had an apprentice, surely this was more in his line of work?

My first meeting with Daniel, as the shop steward, was brisk and to the point but very polite and respectful. I stated that the job was dangerous and not a duty for drivers. He replied that the job had always been done by them and always would be.

It seemed like a stalemate. Not wanting to be too confrontational, I told him the situation as it now stood. No driver was going to do it again. If asked, they would refuse, and the way Daniel handled that refusal would either defuse the situation or escalate it.

It escalated it. I was in Suffolk collecting spuds, when a driver refused and was sent home. Only come back if you will do it, he was told.

I arrived back at the yard, later that afternoon, to scenes of rebellion. Daniel refused to back down, we were told to just get on with our jobs.

A union meeting decided our next moves. Tomorrow was going to sort this problem, one way or the other.

Next morning everybody arrived at work on time. The lorries were loaded one by one. But they didn't pull out, they parked up in the yard. Daniel arrived just after 7am to the sound of his phones ringing, customers wanting to know where their orders were.

Daniel demanded we pull out, or we'd all be sacked. I demanded that the guy sent home was reinstated without loss of pay, and the practice of steam-cleaning be stopped.

He said we'd sort it later, just get the orders out. I replied from a position of strength. No!

By now all the trucks had loaded and the drivers were just sat there waiting.

Eventually Daniel saw sense and conceded, he agreed to end the practice and reinstate the driver. The boys got out on the road and the garage was put firmly in its place.

Things were a bit frosty for a while, but the victory certainly gave the drivers a sense of belonging and respect.

At the end of that long hot summer, I began to get itchy feet. A couple of years hand-balling spuds, and all that roping and sheeting began to take its toll. The thought of another winter out in the fields had my bones creaking. I hadn't driven artics now for a couple of years and I worried I might lose some of my hard-learnt skills (reversing not being one of them). And seeing some of those modern sleeper cabbed units in the transport cafés made me long for a bit of tramping again. My time with Robert's of Knighton had given me the bug. Setting off on a Monday not knowing when and how you would return. New routes, new places. This desire for places new has been with me all my working life.

I got a tip-off from a friend, of a job hauling timber out of Shoreham Docks. At the interview, the boss, Jack Muggeridge, ask me if I'd done timber before.

'Yeah, no problem,' I lied.

The only timber I'd carried was pallets! But I thought: no lie, no job. You may well remember I'd done the same at Robert's, about roping and sheeting. I never learn, do I?

Now I've worked for some crooks in my time, but Jack certainly takes some beating. He was the sort of guy who never looked you in the eye, always over your shoulder, like he was talking to someone else.

He ran Westline Haulage, a small firm, with just one artic, and three Magirus Deutz eight-legger tippers. The yard sat sandwiched between two der-elict warehouses on a small industrial site. My van was the only vehicle in the yard that was taxed.

One of the reasons I went after the job, was the truck. A beautiful DAF 2800 DKX. A real good-looker and one of the big trucks of the day. And my first sleeper cab – no more sleeping across the seats for me. And it had fitted curtains! What luxury! Unfortunately it had a reverse 16-speed gearbox. You moved it away from you, not towards you.

When I pulled the DAF into Shoreham Docks on my first day, I already knew I was out of my depth. I had half a dozen dogs and chains on the flatbed trailer. I'd never even seen them before, let alone knew how to use them. I loaded up huge packs of cut timber for Tunbridge Wells.

'How do you want them, Drive?' asked the loader on his huge fork lift.

'I dunno,' I replied. 'Just whack it on.' And he did, ten packs, over 15' high.

I drove gingerly to the side of the dock and dragged my chains off the trailer. What now? I stood there like an idiot, scratching my head. They were far too heavy to throw over the load. And what way up do the dogs go? Was six going to be enough? I didn't have a bloody clue.

Occasionally on your path through life you meet up with someone who shows care and compassion to their fellow beings. People who help you out because they can, not because they benefit, or feel obliged to.

On that day I met Howard Charman. An owner-driver, not a company man, he'd already loaded and chained down. He could have so easily driven by and smirked at the sight of me looking so lost, as several drivers already had. But no, he parked up and came over to help. He helped me get the chains up over the load, advising me to always put them on the last pack of timber to be lifted up. Then he showed me how to use the dogs, keeping my head out of the way as they have a nasty habit of springing back. And always put the dogs on the offside, so you can keep an eye on them in your mirror. He also told me I should stop and tighten up the chains ten minutes down the road, once the load had settled. He was a saint and I could have kissed him. A gentle, cool guy. Everybody needs a Howard in their lives sometimes.

Over the few months I worked there, he was always on hand to offer help and advice. Whenever I saw him in the docks, he was always helping someone out, folding up sheets or chaining down.

He also warned me about Jack. He was a fly-by night, he said, running up debts then doing a runner. Several owner drivers in the docks were owed wages.

The DAF was the best truck I'd driven. I struggled a bit with the gearbox for a while, but she was a great truck to drive. And nights out were like staying in a hotel compared to some of the wagons I'd slept in. I'd wake up, lean over and put the kettle on, then snooze as it boiled.

One of the first runs I did for Jack was Ford tractors, from Brentford in Essex, to Southampton Docks for export. The 40' flatbed trailer I pulled was a rotting ancient relic with maybe just legal tyres. At the plant, I backed onto a loading bay and the first tractor was driven on. As it reached the headboard, a rear wheel went right through the floor. Fortunately a cross member stopped it going too far. They tried to reverse it out, but the decaying trailer floor just broke up even more. Reluctantly they put another tractor on, gave me the paperwork and told me never to come back. At the docks, the rear tractor had to pull off the first one, resulting in even more damage to the trailer floor. They too told me never to come back.

Jack didn't seem too bothered, and the next day a new hire trailer was in the yard.

He started off giving me cash for the fuel. Twice in two weeks I ran dry on the side of the road and he had to come out with a few cans. Then I began arriving in the morning to find the DAF with a full tank of diesel. Jack was filling her up after I'd gone home.

One night I forgot to take my tacho out. It showed that he'd taken her out at midnight and driven no more than 100 yards or so, before stopping for about 20 minutes. Then he returned to the yard, with a full tank. This happened most nights. The tipper drivers confirmed that their trucks were also miraculously filled overnight, but warned against asking too many questions.

It later transpired that Jack had arranged for a tanker full of diesel to be delivered to one of the empty factories on the estate. When the tanker arrived at the site, Jack was stood there to sign for it, Billy Bollocks. By the time the fuel company sussed they'd been duped, the diesel was all gone.

There were so many stories about Jack. He wasn't a legend, or anything like it, because nobody really liked him. In fact he had a lot of enemies. Owner-drivers pulling out of Shoreham hated him because he undercut them by having no tax on his wagons, dubious insurance and no overheads to pay. Suppliers hated him because he ran up huge bills in various names and always defaulted.

Jack walked with a pronounced limp, the result many said, of a broken leg, suffered when he ripped off the wrong guy.

Another tale – and I believe there is some semblance of truth in this – was when he sold a 'reconditioned' gearbox to a local continental haulier. When the box blew up halfway to Spain on its first trip, Jack was unrepentant. 'Your bloody driver,' he squealed. 'Nothing wrong with the box.' Somebody obviously didn't agree. One night, not long afterwards, the windscreens and lights on every one of Jack's wagons in the yard were smashed.

Rumours of pursuing creditors and looming bankruptcy abounded. The DAF was repossessed and I ended up driving a Maggie. Then, once the taxman got involved, everyone knew a runner was imminent. That's the way Jack handled such things.

So I quit, moved on. Just three weeks later Jack also

moved on, disappearing into the night. His drivers never saw their last wage packet.

Many years later I bumped into Jack at the Avonmouth Services. He was walking away from me, but his limp gave him away.

'Hiya Jack!' I called out, as I caught up with him.

He nearly shit himself! He must have thought I was one of his creditors! Over a cup of tea he told me how well he'd been doing. During the miners' strike he'd been running oil over the picket lines in South Wales.

'Made a fortune,' he declared. 'You should have stuck with me son.'

'No I couldn't do that,' I told him. 'I could never be a scab!'

'You worry too much about other people,' he laughed. 'Me? I don't give a shit about anyone!'

Just one Howard Charman is worth a thousand lowlifes like Jack Muggeridge.

My next job was again with a legendary motor, this time a Ford Transcontinental pulling a bulk tipper. Although you never really appreciate a truck until it's gone, I knew straight away she was something different. She was one of the biggest trucks around, really high up and with a spacious sleeper cab. Other drivers wanted to sit behind the wheel and take photos. She also had air horns – a real deep bull horn sound. What a great way to make your presence known. You could make pedestrians jump out of their skins at 50 yards!

A regular run was grain from Chichester to Cardiff, then a load of clinker out of Barry power station back to Ford, near Littlehampton. One day's work! She just flew, she had too. On the M4 I regularly hit 80mph!

I would often take my two eldest, Justin and Daniel, along with me. They just loved it. On a great sound system, we'd listen to Bob Dylan's *Infidels* as we ate up the miles (cassettes, remember them?). The kids have fond memories of those days. I went out with my dad in his truck. Back then, truck drivers' kids were envied by their mates because they got to go all over the country with their dads during the school holidays. Not possible these days, with health and safety and all – most firms now wouldn't let you on the premises with kids on board.

Being so high up in the Transconti gave you lots of advantages. In country lanes you could see over hedges and seemingly around corners, and your view of the road ahead surpassed that of most other road-users. But it could also have deadly disadvantages.

One day while passing through Worthing, a woman stood at a zebra crossing started to cross as I approached. I hit the Jake brake and the truck slowed right down. By the time I arrived at the crossing I was down to a crawl and the woman was safely across. My foot came off the Jake brake and I went to hit the accelerator. Then out the corner of my eye I saw a small movement! My foot immediately went to the brake instead and we stopped dead. The truck stalled and my gear shot off the bunk into the passenger well.

And across the front of the truck rode a young woman on a bike! The movement I'd seen in the very corner of the windscreen was the bobble on top of her woolly hat! She'd been riding on the pavement and suddenly rode out onto the crossing. If she hadn't have been wearing that hat I would not have seen her, and I would have hit her. No doubt at all in my mind, the consequences would have been fatal. I still to this day shudder at the memory.

Whenever I see a truck with 'Been Everywhere' flags, nameplates, footie scarves, bullfighting spears or any other such crap that these so-called professional drivers decorate their screen with, I think of the girl in the bobble hat. When, as an agency driver, I'm in other people's trucks, I always remove any such vision-blockers and put them away in a locker. One guy even had a pair of panties hanging up in his screen! These wagons wouldn't pass an MOT with that shit in the screen so why are they allowed to get away with it? Over the water, everybody's at it. The French are the worst culprits of them all. Sometimes you can't even see the driver through all the flags. Then there's the TVs, coffee-makers and laptops cluttering up the dash. The Spanish truckers like to collect cuddly toys and fill their dashboard with them. Ahh!

Now, here in the UK, there is this 'really cool' fad of half-closing the side curtains. This means you can't see the driver and – guess what – he probably can't see you! There's enough blind spots on a truck without creating any more. These people aren't professional drivers at all, they're just playing at it.

Unfortunately the bulker job didn't last long as the firm went bust and the Transconti was repossessed. These were the early Thatcher years. Times was 'ard, as they say. Jobs were few and decent wages a thing of the past.

7

Back on the Timber

I got lucky and thanks to my timber 'experience' with Jack, I managed to blag a job with SPS. Southern Port Services. They ran Scania 81s and 110s, with a few Seddon Atkinsons, out of the Basin in Portslade.

Many of the runs I did were carrying tropical timber, coming off the boats from Malaya and Thailand. These huge trunks often overhung the trailer by the then legal maximum of ten feet. You had to nail marker boards on them.

So if these were only the middle section of these trees, how big were they before they cut them down? I carried one single 40' trunk that weighed more than 18 tons. This may have been over 30 years ago, but people knew then the damage it was doing to our planet. Documentaries on the TV kept telling us that these trees were the planet's lungs, so why were we doing it? Why are we still doing it? Most of the trees I carried went to be stored for investors. They were traded as a commodity, making money for people who never even saw them or the hole they left behind in the forest.

The miners' strike eventually reached Portslade. Thatcher had declared war on the unions, and the battle came to the Basin when rumours of a Spanish coal-ship's imminent arrival brought several carloads of Kent miners to form a picket line. They were proud guys, fighting for their jobs and their way of life. They had no problem with us going about our jobs. The only problems would be if and when the coal–boat arrived.

In our breaks, we spoke with them in the canteen. Most people supported them all the way. Thatcher had decided to destroy the coal industry and with it the dreaded unions. These guys had already been on strike for several months and were struggling to feed their families. Regular collections were quite generous and drivers often bought the miners their breakfast in the dock canteen.

The local police were pretty cool about it all. Just keeping a presence, they had a good relationship with the strikers. All this changed when a boat appeared off the coast. We couldn't see it but we knew it was there when busloads of London bobbies and the country's media descended on the Basin.

The new police were very aggressive with everybody. We drivers were ordered away and the pickets got short shrift and were bundled out of the way. One picket managed to climb a crane, amid cheers from all those watching. The ship never attempted to dock. The pickets saw it as a victory, they'd kept it away.

I'd begun using my first CB. With the handle of 'Freebird', I copied breakers all over the country. I found it great for getting directions. You could always find someone to take you right in to your tip. All pre-satnav. And on lonely nights out, you could chat away with other breakers all night long on the issues of the day. Usually politics, football and women, although not necessarily in that order.

The Scanni 81s had a tiny cab with a fold-down bunk that was awkward to put away in the morning with all your gear on top of it. After a while I took over a Seddon Atkinson that had a proper fixed bunk.

Great! But it also had an Eaton Twin Splitter. I hated that bloody box, so temperamental, sometimes it went into the selected gear, other times no bloody chance. Perhaps it was me that was temperamental. My reversing was improving, I no longer lost sleep worrying about the next day's delivery.

Early one morning I pulled over into a lay-by to check my chains. I'd only run for half an hour or so, but the load needed to settle before you could truly tighten it down. I tightened up the chains, one by one. The last one was being difficult. As I pushed down on the lever of the dog, it suddenly sprang back up, whacking me under the chin so hard I bit my tongue. I let go of the dog and it fell, cracking my head and knocking my glasses clean off. I fell to my knees then rolled over onto the ground.

Blood ran down my forehead and into my eyes. I spat out blood, my tongue throbbed. With blurred vision I scrambled about, looking for my glasses. My white T-shirt became sort of red tie-dye.

I took a few moments to gather my thoughts, then removed my T-shirt and wiped my face. Getting my water container from the cab, I rinsed out my mouth and splashed my cut head.

When the bleeding eventually stopped, I sorted out the dog and chain, although loosely, and drove on to my tip. There I managed to get some treatment from their first aid guy. He said it wasn't as bad as it looked. I begged to differ and he called me a wimp!

It was while I was working for SPS that I did my first van conversion. I bought an old Trannie van and put windows in the sides, using back windows from a couple of wrecked Scanni 81s. Written-off trucks were often just parked in some corner of the yard and forgotten back then. The bunks from the same trucks fitted perfectly, one over each wheel arch.

That summer we did a tour of Wales in the van, visiting old friends and camping on the Brecon Beacons. Me and Jo slept in the van and the kids shared a tent. Good times.

First thing in the morning, before you pull out onto the road, you do a walk-around. While the engine is warming up, you do a quick trip around your wagon to check everything is OK. Tyres, lights, indicators, that sort of thing. The importance of the walk-around was underlined by an incident that cost an owner-driver a lot of money. There was no tick list or legal requirement back then, and a quick walk-around was often a lot quicker if it was raining. This particular Monday morning it was chucking it down. As the guy hit the first roundabout on the A259, just a few hundred yards from the dock gates, his nearside drive wheels parted company with the hub and rolled down the road. The truck keeled over and his load of timber came off, blocking the roundabout for most of the day. A crane was needed to sort it all out.

When the investigating police went back to the yard, they found his wheel studs sat on the ground where he'd been parked up overnight. Apparently someone had attempted to steal his wheels during the night. If he had done a walk-around he would have noticed and saved himself a massive fine and recovery bill.

When work dried up at the Basin I was 'last in so first out'. No hard feelings. Bastards!

Within a couple of weeks I got lucky and signed on for Shoreham Haulage. A two-truck outfit running out of Shoreham docks, mostly subbying for Shoreham Transport. My wagon was another DAF 2800. Although not as high-spec as Jack's, it was still a great motor with a good spacious sleeper cab. My days of sleeping across the seats seemed a long way off. And it had a proper gearbox.

But the job wasn't as they said. Promises of lots of work and good load bonuses never materialised, I was at odds with the transport manager and began looking elsewhere for work.

One Monday morning I loaded up with a load of kiln-dried timber, bound for Yeovil. Fully sheeted up with a fly sheet on top. A hard way to start a week.

The trailer brakes had been locking up for a while, so before I pulled out of the yard I called into the garage to get them looked at. There was no defect system then, if you had a problem you went direct to the mechanics. The young mechanic was busy. I asked him to have a quick look.

'Can't you look?' he asked. 'Most drivers do their own brakes around here.'

Well most drivers here were owner-drivers and knew what they were doing. I wasn't and I didn't. When I said I wouldn't be leaving the yard unless they were checked, he threw his toys out of the pram. Then he crawled under the trailer and checked the brakes.

'I've wound them off a bit,' he spat, as he wriggled back out.

In most of the jobs I've had, there has always been a tender relationship between the mechanics and the drivers.

'You wreck 'em, we fix 'em,' says the mechanic.

'You fix 'em, then we break down,' says the driver.

After a quick brake test in the yard, all seemed well and I set off for Devon. The M27 wasn't even thought of then, so the route was down the A27. A hard and slow run past Bognor Regis and Chichester, then a slow crawl through Emsworth.

Approaching the last roundabout before Portsmouth, a car shot past me and dived in at the last moment, narrowly missing the bollards and me. I cursed him as I followed him around the roundabout and off again. I was working up through the gears when suddenly he stopped dead. No warning, no indicators, he just stopped. On the opposite side of the road was a garage. That's why he'd stopped, but a car coming the other way had prevented him from shooting in. I hit the brakes and she juddered. The tractor brakes bit, but locked up and skidded. The trailer brakes didn't happen at all, and with an almighty crashing noise, she jack-knifed across the road. I was flung across the seats, my left hand still clasping the steering wheel. I closed my eyes as my door window shattered and covered me in glass.

Once the noise and movement had stopped I slowly opened my eyes. I was laid across the passenger seat, trembling but unhurt. I could see timber through the side window. Out of the screen, I was looking down the length of the trailer. The cab was embedded in the side of the trailer. The truck was just inches from a wall, and the car coming the other way had ploughed into my wheel.

To say I was shaken was an understatement. The police arrived and the car driver admitted he hadn't indicated or even known I was behind him. He'd meant to drive straight into the petrol station, but had hesitated when he saw the car coming the other way. My tacho showed I was only doing 18mph, but you wouldn't have believed it with the damage done.

After reluctantly exchanging insurance details, the driver went on his way. The police weren't really interested in anything else apart from getting the road reopened. The damaged car called for a tow and, fortunately for me, no cables or airlines had snapped, so I managed to un-mangle her and straighten up. The damage wasn't enough to have her taken off the road (although it certainly would these days), so after sweeping out the glass and taping the wing mirror glass back in, I drove slowly back to the yard.

After a load of abuse from the boss, I got a senior mechanic to look at the trailer brakes. They had been wound right off on one axle! When the tractor brakes came on, the trailer just ploughed on, pushing her to one side. A typical jack-knife.

The offending mechanic denied it all of course, I must have wound them off myself, he said. Thankfully no one believed him.

I was almost heart-broken, trashing such a beautiful truck. So that was the end of that job. Truck off the road and so was I.

8

The Lucking Years: Well Done Son!

TOWARDS the end of the 80s I started working for GH Lucking, the theatrical removers. Based in Ashington, West Sussex, they were just a few minutes' drive from my home on the south coast. At the time they were the UK's biggest hauliers in the entertainment business.

The fleet consisted of 25 DAF 2100 tractor units, 65 trailers and more than 20 various rigids and vans. The distinctive livery and Olde English sign writing made them easily recognisable in the days of plain drab paint jobs.

'Keep on trucking with GH Lucking' was the company logo. Each truck bore the name of a theatre. Gordon Lucking swore that this is where Eddie Stobart got the idea.

Our main task was moving touring shows, musicals, comedies and plays, around the country from one theatre to another. We'd take a show out of a theatre (get-out) on a Saturday night and put it in another theatre (get-in) on the Sunday or Monday. Then during the week we could be working for the BBC, props and location work, or delivering washing machines, car parts or polystyrene tiles. Anything that would fit in the 14'-high single-axle trailers.

I'll start by saying that this job was probably one of the best of my career. I stayed for seven years. Decent trucks, plenty of overtime and frequent trips over the water. The work was hard, long hours off the card and only a couple of days off every fortnight or so, but it was a great job.

Gordon Lucking on the other hand, was the biggest bastard I have ever worked for. A big, mean, arrogant SOB. His persistent use of that nasty 'C' word, in practicably every conversation you ever had, sums up the man completely.

The firm's attitude to their drivers was very much 'if you don't like it you can just fuck off!' Work was still hard to come by, and it's amazing what you put up with just to stay in work. Let me give you an example of their man-management skills. I was stood in the office one day when Peter Crouch, the transport manager, was speaking to a driver over the phone.

'Well done, son,' he said. 'Yes, you've really helped us out. Well done. I wish we had a few more drivers like you on the firm. We need drivers like you who will go that bit extra, well done son!'

When Crouchy put down the phone he turned to Gordon and said, 'What a c★★★!'

The whole office erupted in laughter.

They knew I was stood there, but it didn't matter. Us drivers were just that, c★★★s!

When work was slack, Gordon would sack a driver or two on very flimsy grounds. Then later, when work picked up again, he'd call them up. Laughing and joking, he'd say he'd forgiven them, water under the bridge and all that. Want your job back? Most said yes. The Thatcher years were hitting all workers hard. If you had a job, you held on to it. You took shit that would be hard to believe these days.

But once you got up the road, touring, it was a great job. Most of the shows we moved around the country toured for months at a time. Once a driver had established himself on the show with a few get-outs, he would usually stay with it for the whole run. It would become his show. Second and third drivers, when necessary with big shows, would come and go, but there would always be the same lead driver.

My first get-out was *A Chorus Line*, out of the King's Theatre in Southsea and into the Royal at Nottingham. It was a one artic and one rigid move. I had the rigid, a huge DAF Pantechnicon, and Plank Head, the lead driver, had the 2100. His real name was Dave, but he answered to his nickname, apparently gained when a load of scenery fell on his head, squashing it flat.

Most of the drivers had nicknames used by all, management and drivers alike. Most were self-explanatory, such as Plymouth Pete, Bobby Bullshit and Gay Lord. Then there was Bluey and Wayne Kerr, Bonking Bob and Jesse James, Manchester Pete and the Vicar. I often wonder where they all are now.

I parked up outside King's late one Saturday afternoon. Plank Head took me backstage to meet the stage manager, who would be coordinating the move. Then I watched the show from backstage before the get out began at about 11pm.

The loading was orderly and constant. Plank Head told the crew what he wanted, and in what order. Huge scenery flats were carried out and slid into the trailer, where Plank Head would tie them off. Then the electrics came out and went onto my motor, flight cases, loose lights and cables. My loading was complete when all the costumes were on board. Plank Head told me to take off. He had to wait for the flooring, but he would meet up with me at Leicester Forest Services on the M1.

By now it was 3am. My tacho went in and off I went. Apart from being absolutely knackered, all went well and after meeting up with Plank Head, we ran into Nottingham. Arriving at the theatre by 8am, we managed an hour or two's kip before the get-in began at 10.30am. It was a slow and laborious fit-up. Offloading a few bits here, then nothing for an hour or two, then a bit more.

I wasn't empty until 5pm, then a call to the office had me running back down to London to help with a Monday morning get-in at Shaftesbury Avenue.

And that was more or less the pattern of my weekends for the next seven years or so, and I loved it most of the time. I rarely booked less than 90 hours a week, and seven nights out was the norm, until you had a couple of days off. On the sacred seventh day, when you were legally required to take a break, you'd end up off the card doing a get-in or at a warehouse somewhere, loading scenery on and off trailers,

Getting time off was very hit or miss. You could book your days off weeks ahead, but if a job came up it didn't matter.

'Yeah, of course you can come home, but don't bother coming back!'

The logistics of moving 20 to 30 shows every weekend is mind boggling. Empty trailers from the get-ins would be left close by, in lay-bys or lorry parks, ready for the next get-out. Others would be run in close towards the end of the week. On Saturdays, there couldn't have been a town anywhere in the UK without a GH Lucking trailer parked close by.

On long hauls between theatres, some trailers would be run part-way and dropped. Drivers from other shows, with Monday get-ins, would then pick them up and take them on.

Now you can say what you like about Peter Crouch's man-management skills, (and I do) but you could not knock his ability to run the transport office.

Any problem, he'd have a solution. Day or night. He was doing 24/7 before the Yanks invented the term. He'd know where every trailer in the fleet was at any given time, off the top of his head. Tractor broken down? He'd pull drivers off a show to cover, then replace them with drivers off another show and so on down the line until everything was covered. I don't believe he ever slept at weekends.

He had the catchphrase of 'Well done son, well done'. Every phone conversation ended with those words. Even if you called in on your day off, it would be 'Well done son, well done.'

Gordon and Crouchy played 'good cop, bad cop' all the time. Gordon would scream and shout at you, threatening to end your miserable little life, and Crouchy would calm you down and talk you out of quitting, especially if a weekend was looming.

At Christmas we'd get a bonus, which was usually handed out at the annual Christmas dinner. Gordon would mingle with us drivers at the bar and slip us an envelope. Some would get a good bonus, others would get shit ones. One year I didn't get one at all. There was never an explanation. I think he enjoyed playing these little games. It set the drivers against each other. Divide and rule.

There was a driver, nicknamed the Vicar, who kept a crucifix stuck on his steering wheel, and carried a Bible in his cab. No problem with that, as they say, whatever floats your boat. But he had a nasty habit of preaching, and on a hard get-out, late into the night, the last thing you wanted to know when you had a problem is what Jesus would have done.

Jesus didn't look over him very much though. He drove away from the yard fuel pump one day with the nozzle still plugged into his fuel tank. It ripped the hose right out and sprayed diesel everywhere.

Gordon gave him two options. Pay for the damage with weekly wage deductions or leave now with no final pay packet. He stayed, and by the grace of God, paid off the damage.

I stayed on a rigid for a few months, doing a lot of prop collections for the BBC during the week and then various shows at the weekends.

In Shepherd's Bush, just across the road from the BBC, was Silver Road Lorry Park. A lorry park in London? I hear you say. Yes, but of course it's not there now. Westfield Shopping Centre now stands where I and dozens of other Lucking's drivers once used to lay our weary heads. It was a great place for meeting up with your colleagues and whinging about Gordon. Catching up on all the news about who'd quit and who'd been sacked. It had a café that closed at 8pm, then a butty van that stayed open most of the night, which was great as most of us never pulled in until the early hours.

Most of the Lucking's boys showered and ate at the Television Centre. A tractor unit could fit five or six drivers in and we'd just drive over the road and straight in. Security knew most of us and they just waved us through.

Sometimes you'd meet up with some of the 'stars'. I met Les Dawson and the Roly Poly Girls one night in a green room. What a nice genuine guy he was – happy to chat to a couple of truckers and have a good laugh.

On a couple of occasions, some of us managed to sneak into the *Top of the Pops* studio. High up in the top of the studio is an observation room, like a glass turret. You gained access from the top floor of the Centre. I can remember seeing Slade! Wow! Thankfully, security at the Beeb has tightened up a lot since I was last there.

With the Beeb, you always carried a porter with you when you went out around the prop houses in Park Royal, collecting and returning props from filming. The porter's job was to check everything was correct, sign the paperwork and stow the props on the truck. I just drove.

I'd often get out and wander around these huge warehouses. One was full of medieval props: coats of armour, carts, cotton bales and costumes. Another specialised in the 50s and 60s: avant-garde furniture, jukeboxes and all the paraphernalia that went with that era. There was also a prop house full of toys, ancient and modern, and one filled with three-piece suites.

One day we picked up a huge bubble-wrapped glass swan, a two-man lift. The Beeb had used it on-set the previous week, but had to reshoot the scene and wanted the swan back for continuity. It had snowed all day and the roads were quite icy. So was the pavement. On our way down the A40, heading back to the Beeb, we saw an elderly gentleman slip over on the pavement. He went down with a wallop, striking his head on the pavement. I quickly anchored up, and we both jumped out to help him.

He seemed shaken, but OK apart from a slight graze. Unfortunately he stank of booze, and he was clearly not looking after himself. He told us where he lived, just a few doors away, and we helped him to his door. Inside, the place was a tip. Piles of newspapers and clothes littered the floor, empty beer cans lay everywhere. He assured us he was all right, but our offer of calling an ambulance seemed to terrify him. He started yelling and swearing at us, telling us to get out of his home.

As we left I knocked on his neighbour's door, I felt I needed to let someone know about it — after all, he had struck his head. I knocked several times before the door slowly opened, fastened with a chain. An elderly woman peered out.

I explained what had happened and asked if she would check on him.

'Piss off!' she screamed, slamming the door shut.

In spite of such hostility, I felt we'd done our good deed for the day. My porter thought we should have left him on the pavement.

A short while later, I had to agree with him. When we opened the back doors, back at the Beeb, we found that the swan, untethered, had lost out to gravity when I slammed on the anchors. It had travelled up the wagon and, smashing into the headboard, had disintegrated into a million pieces.

There was hell to pay. Gordon became involved, and I got a right bollocking. The BBC was god: never upset the BBC. The porter's remit was to load and secure the props, so he was suspended for a couple of weeks and Gordon's insurance didn't have to pay up.

So remember children, a good deed doesn't always earn you brownie points.

Some of the best jobs going were on TV film location. I sat in a lorry park in Whitby for more than two weeks with a lorry-load of props and filming equipment. They were filming *Vanity Fair* and on filming days all meals were free from the chuck wagon. You just got your eight hours' pay and nights out, but it was a good break from the weekend get-outs. You just had to be on-hand to open up the truck as and when required.

Another break we got was when we took a show to Aberdeen. If it was only there for the week, we'd stay with it. It was too far away to leave, then travel back. Again we'd get our eight hours a day and nights out.

Aberdeen is a beautiful city. In the spring, beds of daffodils line all the verges for miles around. It's called the Granite City, and those granite buildings really do glisten and sparkle on a sunny day.

It was about this time that I started to write, and a week off in Aberdeen was a good time to get stuck in. I was always good at telling stories, to my school pals, then the kids and occasionally to my workmates.

My first attempt was about a trucker (naturally) who picks up two runaway kids in his wagon. After a bit of research and study, I changed it from a novel to a TV screen play. Called *Becky* after one of the kids, I submitted it to the BBC and got a good response. I had a couple of meetings with a scriptwriter, but then things faded away and a change of personnel in that department was the end of *Becky*.

I didn't despair (oh yes I did) and started to attempt a few short stories and radio plays. Again without much success. But I ploughed on.

I moved on up to artics, my first tractor unit was named 'Haymarket'. I did a couple of runs down to Disneyland Paris, back then just a few warehouses in a large field. The sets and scenery came out of a set-builder's in Brixton.

It was my first real trip abroad. I was amazed at the standard of truckstops and the ease of driving on the autoroutes. They were mostly two-lane, but you could always make good progress, really eating up the Ks.

But the Périphérique, the Paris ring road, scared the shit out of me! Crazy, crazy drivers seemingly intent on destruction, one hand on the horn, the other wildly gesticulating out of the window.

I followed that by taking Wayne Sleep to Amsterdam, a three-truck move over three days. After an all-night get-out, we ran it to Leeds. A terrible run, chasing the clock and battling to stay awake.

There was a heavy drinking culture among the

drivers. I'd had my drinking days in the Merchant Navy, there wasn't much else to do when you're at sea for eight or nine weeks at a time. Now I wasn't bothered about it at all: a half before a get-out was it for me. But my colleagues would often get legless, then after the get-out would drive off to the next venue. Lucking's were certainly aware of the problem, as several tour managers complained and had drivers removed from their shows. But they just got another tour and carried on.

My reversing skills certainly improved. They had to. I was now running trailers in and out of the West End. And not just any trailer. With a single axle, the 40-footers were a pig to reverse. Cutting in so quick, you had to anticipate so bloody far ahead. And they weren't much better going forward. Boy did they cut corners, you needed a lot of space on roundabouts.

In the back streets of Soho on a Saturday night, a badly parked car on a T-junction would cause chaos over the whole area as traffic backed up around you. We had a phone number to call in such circumstances. Within a short space of time a tow truck would appear and remove the obstruction. You get a completely different take on traffic wardens. We loved them! Without them, the West End of London would be gridlocked every night of the week.

You really had to be careful when driving in the West End. Any time of the day and night, people are everywhere. Pedestrian crossings often have an unending flow of people whose minds are not always alert to what's happening around them. They often do stupid things. On one occasion a young couple got fed up waiting for me to clear a T-junction and actually ducked under the trailer! The distance between the landing legs and the tractor mud guards is so small that any movement from me would have seen them under the wheels. There was a need to be alert at all times – and not only when you're on the move.

Sat outside the dock door at the Alhambra Theatre in Bradford one Saturday afternoon, waiting to do a get-out, I watched as an immaculately dressed elderly lady walked slowly by. Her style of clothing looked early 50s: tweed jacket, leather handbag and a feathered hat. She suddenly turned and went down a short alley in front of the truck. I knew the alley was dead end, the crew sometimes had a car or two parked down there.

Maybe she'd nipped down there for a pee, I thought, although it seemed unlikely. After four or five minutes, she hadn't emerged and I became a bit concerned. Maybe she'd been taken ill or even been attacked! I climbed out of the cab and went in search of her.

Shock horror! There she was lying sparko on the ground.

I ran to her and rolled her over onto her side. She seemed to be breathing OK, so I ran back to the stage door, yelling at them to call an ambulance, quick!

I held her head in my arms, telling her everything would be all right. The police were there in minutes. A policewoman casually ambled over.

'Hallo Mary, you naughty girl, this is the second time this week.'

She pulled her up by her shoulders, sitting her up. Then reaching into her handbag, she pulled out . . . a bottle of methylated spirit!

I was so shocked! Who would have thought such a sweet little old lady could be a meth addict? The WPC explained that Mary lived in a home and kept going AWOL. Hardware shops in the area had her picture, with orders not to sell the stuff to her.

As my mum used to say, you can never judge a book by its cover.

I began to get my own shows. One of the first was Opera 80. They moved every three days so I tended to stay with them all week. They were touring *Rigoletto* and *Cinderella*. Now I'd never been an opera fan, I liked the music but never understood why they always sang in Italian. But once the stage manager had explained the plots, I must say I really enjoyed them and watched them several times over. *Cinderella* was the original opera, not the pantomime. Similar story, but without the pumpkin or 'he's behind you'.

While down in Plymouth with Opera 80, I met a very enterprising young chap. I was just backing into a space in the city's lorry park, when a young boy, maybe 12 or 13 years old, appeared and began watching me back, expertly waving me in. He then came around and leapt up onto my doorstep.

Holding onto the mirror, he said 'Look after your lorry for you, mister? Only a quid.'

'Why does it need looking after?' I asked.

'Someone had their red airline cut here last week,' he stated.

'Why? 'Cos he didn't pay up?'

'No, it's not like that!' he snapped.

I laughed and gave him a quid. He told me he was saving up for a bike. Was this extortion or enterprise? Not a lot of difference in the real world is there?

A few days later I rode solo into Silver Road to pick up a trailer for tipping at the Saatchi Gallery in the West End. The trailer was just sat there, unguarded and available to anyone with a tractor unit. I only say this because when I tipped it later that day, it was full of works of art. Huge 20' × 10' murals and landscapes. But my favourite had to be *15 Elvis Presleys* by Andy Warhol. The gallery porters told me it was worth £15 million!

9

Squeaky Bum Time

WE did many different jobs during the week, in-between our shows. After finishing a weekend get-in, you'd often pick up a trailer in Silver Road with 20 drops of polystyrene tiles, seed trays and fish boxes on board, to be hand-balled off at garden centres, DIY stores and fish farms all over the country. Hard work and chasing the clock all the time. You just had to be empty and back with your show for Saturday night. No excuses. The show must go on!

One of the worst jobs we had was without a doubt the dreaded Pannal to Dagenham run. We'd hand-ball foam car seat cushions, floor to roof, at the Dunlopillo factory in Pannal, just outside Harrogate. It took more than four hours of sweat and toil to load. Then you'd run four hours down the A1 to Ford's car plant at Dagenham and hand-ball the bloody lot off again, another four hours. Kip for a couple of hours, then back up to Pannal and do it all over again. If you upset Gordon you could be on this run for weeks.

I was parked up one Thursday in Buxton, Derbyshire, with Opera 80, when I got the dreaded phone call. Over to Pannal solo, pick up a trailer on-site, load it and do a run. Then take the empty trailer back up to the Alhambra Theatre in Bradford, before returning solo to Buxton for the Saturday night get-out. Quite a tall order, but then it always was.

I was over to Pannal by late afternoon but had to wait while two others finished loading in front of me. The winds had picked up and the trailer was rocking as I sat waiting. I wolfed down a can of beans followed by a tin of milk pudding. Cab cuisine eh? Away from the BBC canteen, I rarely ate proper meals. Vitamin milk drinks and breakfast bars were the easiest to scoff down when you were on the move.

The long slog of loading was made even worse by the horrendous smell of hot foam. It stuck in your nose and made your eyes water. I always got a headache at Pannal.

First you stacked the cushions shoulder high, then you made steps out of the next row, so you could stand to take the first stack up to the roof. Repeat 20 odd times and you were loaded, which took over four bloody long hours!

Then all you had to do was drive. It was gone midnight before I pulled out, with the wind now gusting across the open plains of the A1. With no weight on at all, she did some serious side-shifts that had my bum sticking to the seat.

I was over my driving time by the time I got to Ford's and went straight on the bay to offload. Marching up and down the trailer, arms full of stinking cushions, I'd pass them down off the back onto the bay, from where a guy would pick them up and put them in a large cage. When the cage was full, he'd move it away with a forklift truck, then bring in an empty one. It was slow and laborious, with a lot of waiting around. When he went off for a 30-minute tea break, I crashed out on a pile of the cushions. He had to whistle loudly to wake me up.

By 9am I was tipped and ready to go. Old card out, new card in – that's the way it was. I made myself a cuppa in the cab and ate a tin of cold rice pudding

My tastebuds had been nullified by the fumes, I could have been eating anything.

A cheerful DJ informed me of 70mph gusts and a lorry blown over on the approach to the Dartford Tunnel. Across the lorry park, two drivers fought with the curtains of a tautliner, trying to tie them back. As I pulled out, one of them gave me the thumbs up, wishing me good luck. I felt I might need it, with further radio reports of trees down and fallen power cables. I'd never seen winds like this before, except maybe in the China Seas.

As I made my way along the A13, heading for the North Circular, an advertising hoarding was down across the road. Cars were able to drive around it, but I had to gently drive over it. Enough was enough, she was rocking like good 'un. I pulled over and phoned Crouchy from a payphone.

'Yes son,' he said, from his nice warm office, 'I know it's bad, But I've got to have that trailer in Bradford for tomorrow. Can't do without it. And you've got to get back to Buxton. Just take it steady, well done son!'

Back in the cab I turned off the radio and turned on the CB. The 19 echoed with drivers urging caution. Another truck had gone over north of the tunnel.

On the M11, everyone was down to 25mph. My trailer suddenly slewed into the middle lane. I leaned across the cab, as if to counterbalance it. An approaching tipper truck swerved into the outside lane to avoid me. Next moment, the trailer was on the hard shoulder kicking up dust as the wind changed direction.

Off the M11 and onto the A604. A fridge motor on the other side of the carriageway lay on its side. Breakers confirmed that the driver was OK.

On to the A1 and I was down to 15mph, with the trailer snaking one way then the next. Cars hung back, reluctant to pass me. Suddenly the whole truck seemed to take off and I was thrown across the engine cover. I stab at the brake pedal and we come to a halt sideways across the carriageway, but still on my wheels. I'm soaked in sweat and crapping myself. I straighten up and continue northwards. What else can I do? Parking up in a lay-by isn't an option, I'd be a sitting target for the swirling gusts of hurricane-force winds. Like a ship on the ocean, you feel you have to drive headlong into the storm, keep up the forward motion.

A few miles on and I pull into Kate's Cabin. And there in the lorry park is an artic, laying on its side. The driver had been eating his breakfast in the café and watched out the window as it went over.

It was now 1pm. It had taken me nearly four hours to travel no more than 60 miles.

I phoned Crouchy again. Same old story, he knew it was bad, he had problems all over the country. But he needed that trailer in Bradford. I was the only one, etc., etc. Well done son, well done!

Other drivers coming into the café all had their stories: trees down and trucks over everywhere. This did nothing for my confidence. After an hour or so of soul-searching I reluctantly set off north.

At Stamford, I joined a five-mile tailback as motorists zig-zagged around two fallen trucks. At 5pm, I was approaching the start of the A1M and I pulled over into a sheltered lay-by for a break. I listened to the news. The whole country has been smashed by the storm and a truck driver has been killed on the M4 in South Wales.

As I pulled out again, a breaker on the CB warned me of a truck on top of a car at Ferrybridge, with an eight-mile tailback. I decided to divert onto the M18 and rushed to get there before the tailback reached it. For a while, the winds dropped a little and I made 25mph. Then, as I merged with the M1, the wind became head-on and I was down to 12mph.

Radio Sheffield informed me of 90mph gusts over the Tinsley Viaduct and an ambulance on its side in Leeds. This was news I can do without. My whole body ached. My hands were clamped to the steering wheel and my buttocks were clenched to the seat. I was talking to the truck, steadying her on, telling her everything will be all right. What tales we'd have to tell to the boys in Silver Road!

I decided not to risk the viaduct so I came off the M1 and went around by cutting through Sheffield city centre. All went well and I relaxed a bit as the city's tall buildings give me some respite from the wind. I stopped at a garage and filled her up with diesel, the extra weight might help. I grabbed a quick coffee and pushed on.

By the time I rejoined the M1 at Barnsley, it was gone 9pm and I was exhausted, mentally and physically. Another hour or so and I should be there. The solo run down to Buxton would be a piece of cake; I'd be needing another tacho though.

Just north of Junction 39 is the River Calder. The motorway sweeps across it in a slow left-hand curve. As I approached, I could see the wind blasting across in front of me. Bits of debris hit the cab as I braced myself for the next onslaught. Trees on the riverbank were bent double and I could hear the wind screaming around the juddering cab. Halfway across and the entire truck seemed to leap into the air, then crash back down again. I was thrown sideways across the cab and banged my head on the passenger seat. As the truck levelled out, I sat back up and tried to regain control. She was going broadside, trailer first, towards the central reservation. I quickly checked my mirror to see if the outside lane was clear before yanking on the wheel to straighten her out. I thought I'd made it when another gust hit and over she went. And this time there was no return.

The truck ended up on its side across all three lanes, the driver's door embedded in the central crash barrier. My five-gallon water carrier whacked me on the head and diesel poured into the cab. My foot was trapped underneath the brake pedal and I couldn't hit the stop button, so the engine carried on running and the wheels kept spinning. I calmly knocked her into neutral.

My first thought was 'Fucking hell, I've got away with it! I'm not dead! I don't even hurt that much!'

My second thought was 'Fucking hell, this is the third time I've been over! Is my karma really that bad?'

The screen had cracked and come in. Through a gap I could see cars using the hard shoulder to drive around me, and then drive on! They even wound down their windows to get a better look! It seemed a while before someone eventually stopped and came running over.

The guy panicked a bit, pulling out the screen and yelling, 'Get out! Get out! There's petrol everywhere!'

I heard myself coolly saying, 'It's OK, it's only diesel. Don't worry!'

As the windscreen came out the wind ripped in. A pile of tachos on the dash, along with my stash of Shell coupons, went straight out the cab on a twisting whirlwind.

'My coupons!' I cried. I'd been saving for months for a new alarm clock.

'Eh?' said my would-be rescuer. I'm sure he thought I'd bumped my head and had gone doolally.

I found the stop button before clambering out through the windscreen. I was soaked in diesel, my skin tingled. The wind hit me as I stepped down, blowing me into the central crash barrier. I dropped to my knees using the cab as a windbreak.

The cab was bent over the barrier and a long gash had been ripped into the trailer. I felt even better now, that I'd hardly got a scratch.

The police arrived and I sat, stinking of fuel, in the back of their car as they sorted it all out. They weren't all that friendly, I didn't get the comfort and sympathy that I felt I deserved. I think I was beginning to succumb to shock: the realisation of what's just happened, and reliving those last few moments over and over brings you down from the euphoria of surviving.

The police phoned Crouchy, they needed his permission to right the truck, and an address to send the bill. They asked me if I wanted to talk to him, I declined the offer. If I'd have spoken to him then, I'd have probably been out of a job!

As we waited for the wrecker, I listened in on the police radio to reports of other trucks going over. More fatalities, with structural damage to buildings and power lines down.

When the wrecker finally arrived, some hour and a half after touchdown, the police commandeered a fully freighted truck from the ever-growing queue, and sat it on the hard shoulder as a windbreak from the gusts still blasting up the river. My truck was then unceremoniously dragged to her feet. She shook and shuddered, looking a real sorry sight. The curtains flapping wildly out of the broken windows. I was feeling pangs of guilt for treating her this way.

I assumed she would be towed away, but no, with encouragement from the wrecker driver, I climbed back in through the windscreen. Both doors were too bent to open. I flicked the ignition and to my surprise she fired up. Followed by the wrecker, and with my windbreak running alongside me, I drove her off the motorway at Wakefield and parked her up in the first lay-by we came to. After loading all my diesel-soaked gear into my sleeping bag, the wrecker driver dropped me off at Jo's sister's place in Barnsley, at 3am.

I had a cup of tea and a shower before I phoned Crouchy.

'First things first, Mick,' he said. 'Are you OK?'

'Yes,' I replied. Surprised at his concern.

'Good. Got a pen and piece of paper?'

He then proceeded with what I was to do next. I was to get a taxi back to the truck, drop the trailer in the lay-by and run solo to Walker's of Wakefield's yard in Wakefield. Leave it there, then another taxi to a truck rental in Leeds, where a truck and trailer would be waiting for me. Then on to Bradford, drop the trailer and back to Buxton.

'Erm ... What about my diesel soaked gear?'

He suggested I left it with Heather, my sister-in-law, she could wash it for me. He'd work me back that way in a few days' time so I could collect it.

And that's just what happened. I didn't get home for another five days; then just one day off before I went back up the road.

Legend had it among the Lucking's boys that I phoned up Crouchy and told him I'd broken a wing mirror. Crouchy went mad and demanded to know how. To which I coolly replied, 'The truck fell over on it!' Great story but, sorry boys, not true. Wish it was.

The truck was a write-off, Haymarket was no more. The positive outcome of this sorry saga was I got a new truck. With only a few thousand on the clock, she carried the name 'Old Vic'. People started calling me Vic. Her number plate was D460 OGO, affectionately known as OGO, she was to be my mount for several years. I loved that truck like a cowboy loves his horse. I'd talk to her on long motorway runs, urging her up Shap, guiding her through the back streets, and begging her to make it to the next fuel station when, once again, I'd pushed my luck.

One of the first runs I took her on was up to Aberdeen with *The Pirates of Penzance*. The weather took a turn for the worse as our convoy, two artics and a ridged, hit the A74. I was on the front door as the fog came down and we couldn't see a damn thing. We were debating over the CBs as to whether we should pull over and stop when, up ahead, I saw some yellow hazard lights. It was a slow-moving tipper doing maybe 20mph. I locked on to him and we pushed on for several miles before he suddenly came to a halt. I watched as the driver got out of his cab and came back to speak to me.

'Do you know you've just followed me into the roadworks?' he asked. 'I left the main road a couple of hundred yards back.'

Laugh? We nearly wet ourselves. We'd followed this truck blindly into a set of roadworks! We concluded the fog was really bad, so we pulled to one side and waited a couple of hours till it lifted. The workers made us tea in their hut, as they took the piss something rotten.

10

Stressful Times

IT was about now that cab phones began to appear. Gordon was always ahead of the game as far as new technology was concerned. It started off with one or two trucks on big shows having them first. The size of a briefcase, they were like field telephones. Battery life was short and you always seemed to be in areas where they wouldn't pick up.

Eventually all the motors were fitted out with the early Motorolas. They were great to start with. The long searches for a working phone box, where you could park closer than a hundred yards away, was such a bugbear that we all embraced them. We could call the office from the warmth of the cab on a winter's night. Our wives and partners could call us any time they liked.

And so could Crouchy! And he did! He worked a 24-hour day, therefore so did we. And god help you if you turned the bloody thing off.

It certainly changed things. Everything got a lot more hectic. And sometimes it got to you, it really got to you.

I'd just run *Spitting Image* to Birmingham. The show was written and rehearsed in a studio in east London. We'd pick up the puppets on a 30' trailer on a Tuesday, then run them to Central Television for the filming. The scenes were shot overnight, and then we'd run it back down in the early hours of

Wednesday morning. The show was on the telly that night.

Just as I was heading back to London, maybe two or three in the morning, a call from Crouchy changed it all. Drop the trailer anywhere and run solo north. A trailer was on its way down from Manchester that needed to be in Wimbledon by six o'clock.

I met up with the driver at Hilton Park, on the M6, grabbed his trailer and shot off. An hour or so later, as I hit the M1, Crouchy phoned. He was panicking, the wrong fucking trailer! The driver had brought me down the wrong trailer!

I turned around and shot back. I waited half an hour at Hilton Park before a very worried driver pulled in.

'Will he sack me?' he asked.

'No,' I replied. 'We're far too busy. Just don't go back to the yard for a week or two, Gordon will have calmed down by then.'

Crouchy was phoning me every half hour or so. Big cock up. Big customer very cross. He'd given them the old 'puncture' story.

'Well done, son, well done. How long before you get there?'

I was knackered, I'd run all night, and now I was down to run all day. When I'd tipped this trailer it was back to pick up the *Spitting Image* trailer.

I stopped for a pee and took a coffee back to the cab. I'd missed a call. It was soon ringing again.

'Where the fuck were you?' Crouchy yelled.

'Needed a piss ...'

'Don't you carry a fucking bottle?'

He wasn't happy. Neither was I, the pressure on me was rising. I set off again, slugging my coffee as I went.

The North Circular was shit. As I told Crouchy when he called again.

Somebody dived in in front of me and I whacked the horn and screamed abuse at him. An old dear in front of me didn't seem to notice when the lights went green. Again I was on the horn and flashing my lights. 'Get a move on you silly old cow!'

Bloody phone was ringing again. Guess who?

'I've told him you're half hour away,' he said.

'Why?' I asked. 'I'm not!'

'Yeah, but not much more. Said you'd been held up by an accident.'

'Thought you said a puncture?'

'Yeah that as well. Well done son, well done!'

I cut down the A40, then down Wood Lane past the Silver Road lorry park. That's where I'd be tonight, once I'd done Wimbledon, shot back to Birmingham, then returned *Spitting Image*.

As I approached Hammersmith flyover the phone rang again. I tried to ignore it, a motorcyclist was trying to commit hara-kiri, but my blasting horn couldn't drown out that bloody phone.

'Where are you now, son?'

'Bloody Hammersmith! Leave me alone will you?! I'll call you when I get there!'

As I swung wide around the roundabout, two cars, unable to recognise my flashing amber indicators, cut up my inside as I turned left down Fulham Palace Road.

Screaming out loud, I swung the truck across their path, stopping them dead. Slamming the handbrake on, I leapt raging from the cab.

The first car mounted the pavement and nearly ran me over as it shot around me. The second car followed, the young woman at the wheel waved two fingers through her open window as I yelled out abuse.

I was bloody fuming, I'd lost it. Back in the cab,

I floored the accelerator and shot off. And there, stopped at the next set of lights, was that damn woman! I flipped. Screaming up behind her, I stopped just inches from her boot. Leaping from the cab, I ran to her door, and through her open window, I let rip.

Did she not know what indicators meant? Did she think it was OK to undertake a truck on a roundabout? Did she even have a licence?

Well, that was the gist of what I said, but believe me, and I'm sure you will, it was said with a lot more energy and vigour.

I am ashamed to say I completely lost it. People on the street stopped to stare as I went off on one. She was lent away from me, over the passenger seat, trying to put some distance between us.

Then I noticed the look of fear on her face, a tear was running down her cheek. I stopped my onslaught. What the fuck was I doing? She was crying! She feared violence! I stepped back, holding my hands up, muttering something like, 'don't do it again' and walked back to the cab.

I felt wretched. What the fuck was that all about? That's not how I see myself or, I hope, how others see me. As I followed her down the road and across Putney Bridge, I felt disgusted with my behaviour. I had been a complete and utter arsehole!

I contemplated approaching her at the next lights and apologising. But didn't. It would probably have freaked her out even more. The pressure had got to me. How could I have got so angry? It was only a bloody trailer for fuck's sake!

It took me a long time to recover from this incident, you can so easily lose respect for yourself. But I hope I learnt from it. I've never told anybody else about it, until now.

My karma was pretty shit for a while, deservedly

so. After one particularly long get-out, I ran a panto-mime set to a warehouse in Lincoln. It was on a semi-derelict industrial estate in the middle of nowhere. There I met up with a six-man crew at 10am on a very hot Sunday morning.

The motley crew was made up of a couple of Hell's Angels, a punk, two hippies and a pensioner. They'd been paid four hours double bubble for the job, but their mission was to be finished and in the pub by noon.

My mission was a get-in at Bromley, four hours away, by 6pm.

We were all pretty confident and the boys tore into the set. Now I'd done job and finish before, so I knew what to expect. It can be quite dangerous, so I stood well clear.

Two of them jumped up into the back of the trailer and started to throw the set off. And I mean throw! Thirty-foot flats were slid out and dropped off the back, then manhandled into the warehouse. Sweat poured out, fingers were squashed and heads were cracked. Tempers flared and curses were screamed. I kept well out of the way, just carrying the odd prop or two. Even so, I broke into a nasty sweat. I took off my jacket and wiped my brow.

A flight case crushed the pensioner's foot, but as the guy lay screaming, the others ignored him and pushed on. A cry went up as the last case was thrown into the warehouse.

Noon on the dot and they were locking the doors and climbing into their cars. I declined the offer to follow them to the pub, my day was just beginning.

It took a while to gather up all the ties and straps, and then sweep out the trailer. No hurry, I had plenty of time. I shut the trailer doors and headed to the cab. A cuppa, a ciggie, then off to Bromley.

The cab was locked, security was second nature to me. I reached into my pocket for the keys, then realised that they were in my jacket. And my jacket was in the warehouse! The locked warehouse!

After kicking myself and the warehouse door a few times, I sat on the ground and lent back against the doors. I was in trouble. The middle of nowhere and no phone for miles. What pub did they say they were going to? They didn't.

I walked down the side of the building. Bricks, rubble and brambles made the going very difficult. The windows were all barred, but I found one right at the back of the warehouse that had a small half window, and it was open! The adrenaline rush made me ignore the waist-high brambles and nettles tearing at my flesh as I waded towards my goal. The window ledge was head-high, so I used a breeze block as a step to scramble up. Sticking my head through the tiny window, I saw it was a small disused loo. The pan was directly below me, and was full of a brown goo.

I squeezed one arm through the window then followed with my head, like a contortionist. Pushing off the window ledge with my feet, my bum came up and my head went down. I then managed to get my other arm through and began rocking like a see-saw. Legs and bum out, head and arms in.

I suddenly thought, what if I get stuck? It's Sunday. Nobody about and I'm round the back out of sight anyway. I could be here for days.

At full stretch, I managed to grab hold of the pan. I pulled and kicked, wriggled and swore. It worked, I came crashing down! My head hit the door, face first, and my knee struck the pan as my foot went straight down into the putrid mess.

I struggled to my feet, trying to counter the puke

factor of my soggy foot by concentrating on the pain in my knee, the scraped face and the nettle stings.

Be positive, I told myself, you're in!

I opened the loo door, only to find the doorway blocked by huge scenery flats stacked against it. It was like a wall, I could have cried.

The canvas backed set was laid at an angle to the wall, leaving a two-foot gap at the base. Dropping to my knees, I squeezed down into the gap. It was too tight to crawl so I laid on my side and slithered like a snake through the dust and spider webs. It was really tight, I had to force my way through.

After about ten yards of hard toil, I emerged into the warehouse. Battered and bruised, but triumphant. My soggy foot squelched as I limped over to my jacket and keys.

I let myself out of a fire door, slamming it shut behind me.

I didn't have time to feel sorry for myself (although I did anyway). I didn't even have time for that cup of tea. After changing my socks and shoes, I set off for Bromley.

I began to get more of my own shows and did some serious moves. I took a 60s music show to the Isle of Man, during the annual TT week. How good was that? It was a hell of a trip to get there from Brighton, and I needed a police escort from the ferry to the theatre because the truck was over the legal length for the island. A great week was spent seeing the show most nights, travelling on the steam railway and, of course, watching the racing.

The Grand Opera House in Belfast was on the circuit of most touring shows. Although I'd been there briefly with an agency years back, spending a week there was a little worrying. 'The Troubles' were nightly news on the TV. Bombs on the mainland and an attempt to kill Thatcher in Brighton brought home to you just what people would do to promote their cause. Some of Lucking's drivers were ex-army boys who had done tours there. They refused point-blank to go back.

My first visit was with a comedy play starring Una Stubbs and John Savident, aka Fred Elliott in Coronation Street. Most drivers opted to do the get-in then run back up to the ferry port at Larne. They'd stay the week there, running back into Belfast on the Saturday for the get out.

But there was very little to do in Larne, apart from drinking, and I knew that the theatre laid on a minibus midweek to do a trip out for the cast and crew. So after the get-in, I ran the trailer back up to Larne, dropped it in the docks, then returned to the theatre in the tractor unit, managing to squeeze her into the car park at the rear of the theatre. Some nights I'd hear gunshots in the distance and a couple of times I was woken in the early hours by army helicopters with their search lights lighting up the car park.

But I got my trip on the minibus. Along with Una and John – both such nice people, very relaxed and easy to get on with – we toured the Bushmill's whiskey distillery and then visited Giant's Causeway, an amazing natural wonder.

In the bus I sat up front with the driver, the theatre doorman. When I raised my camera to capture a huge IRA mural on the side of a house, he hissed at me to put it away. And when we got stopped at an armed police check point, he again warned me not to take any pictures. 'Folks around here don't like having their pictures taken.'

He later told me that his brother had been shot dead by the IRA. His killer had been jailed but was now out. He saw him every day on his way to work, but had to live with it. He didn't want revenge, he just wanted it to stop.

The soldiers I saw on patrol were just a few years older than my own kids. With guns at the ready, they walked the streets.

The shows promoters were Mobil, the oil people, and they arranged a meet and greet dinner in all the towns that the show appeared. Because I was staying over, I got an invite. It was a posh affair in the Europa Hotel, right alongside the theatre. The restaurant was up on the sixth floor. I sat at a table with the show's lighting boss and five guests of the promoters. They were obviously disappointed not to get Una or John, but they got over it and we had a great night.

One of the guests was a teacher. She asked me if I knew that the Europa Hotel was the most blown up building in Belfast. I tried to stay cool, but she guessed straight away that I didn't. I convinced myself she was joking.

She taught at a mixed school, one of the few in the Province. She told me that the playground reflected the streets. Bitterness and hatred on all sides. Everybody had a relative or friend who had died in the struggle, and every death recruited more to the cause.

I ran to Belfast quite a few times after that. Then one year a car bomb went off outside the Grand Opera House. The car's engine flew up into the air and crashed down through the roof and onto the stage. Thankfully no one was seriously hurt, but the theatre remained closed for many months. Less than 18 months after it reopened, another bomb blast forced it to close again. Both of the bombs were targeting the Europa Hotel next door.

11

Life on the Road

MY being away from home for so long could often be hard on the rest of the family. Jo had three growing kids and a household to run. Being on the end of a phone sometimes wasn't enough. My rare rest days were always midweek, and then rarely more than two in a row. Not a lot of time for playing with the kids or for time alone with Jo.

If I was passing close by, say within 30 miles or so, Jo would drive out to meet me and stay over. We had our favourite lay-bys. We'd also meet up when my shows took me to Brighton, Crawley or Southampton. Jo would always get up in the trailer and give me a hand with the get-out.

At the Christmas parties, the wives and girlfriends of the drivers called themselves 'truck widows'. They called us 'cab happy' – happiest when we were on the road.

Family crises were hard to bear. I'd run up to Aberdeen late one Sunday only to learn when I got there that my father had died. He'd been ill for quite a while, but it was still a shock. Gordon got me on a flight the next day and I was home within 24 hours of getting the news. I took some time off, and when I returned I thanked Gordon for dealing with it. At times like that you need people to do things for you. I shook his hand.

When I got my next payslip, I saw that Gordon

had deducted the air fare, £50, from my wages. He'd also not paid me from the day I'd flown home. Well what did I expect? I was just annoyed with myself for shaking his hand.

In 1991, I began touring *Chess*, the Tim Rice and ABBA musical. It was a four-truck show, moving every two or three weeks. It was a hell of a get-out, often taking more than 12 hours. Then you'd run one trailer halfway to the next venue, drop it in a lay-by, then back to the theatre and run the next trailer all the way. Other drivers would run in when they'd finished their get outs and help out running the others. Get-ins would begin on the Monday morning. Once the trailers were empty they'd be ferried out to lay-bys and left. Sometimes they'd stay there until the next get-out, others would be taken away by other drivers.

One move we did was from Cardiff to Dublin, with a week off when we got there. I managed to smuggle Jo onto the show and she came out with us. Her visiting mum, Kathleen, was on kid-sitting duties.

The weather was just great and we had a wonderful time, parked up right alongside the River Liffey. We explored the city and strolled in Phoenix Park. One day we all managed to squeeze into a tractor unit and took off up the Wicklow Mountains and had ourselves a picnic.

On the Thursday it rained, so we all sat inside our cabs, just chatting among ourselves on the CBs. Then a local breaker came on, saying he was in the IRA. He knew where we were and he'd be down to see us. Now we all know you get wind ups on the 19, but ... the tone of the man's voice and the viciousness of his threats ended the joy of Dublin for all of us. We were all pretty glad when we finally set off for home.

Later in that year Gordon bought five new DAF 2700s. For the first time ever, he didn't paint them up. No air dams or top hats and just plain white. I was pleased to be given one, but a bit disappointed with the livery. I was also a bit sad to be saying goodbye to OGO. She'd served me well. When I cleared my gear out of her for the last time, I felt pangs of guilt for leaving her. I even said goodbye to my old friend.

In the beginning my new wagon didn't have much luck. I'd only had her for a few weeks when, parked up one night and in my bunk at Lesmahagow truck-stop in Scotland, a trailer change in the middle of the night went horribly wrong.

The noisy changeover, drivers swapping trailers, had already woken me up. It was 3am! And I suppose it wasn't really a surprise to discover that a trucker who didn't give a shit about fellow drivers getting their well-earned kip, wouldn't be too bothered about his handbrake either! Something he only discovered when he wound his trailer legs-up. The whole outfit ran away across the lorry park and smashed a glancing blow across the front of my wagon. I was jolted out of bed as the trailer came to a halt, firmly jammed up against me. I lost my grill, wipers and windscreen. And an open carton of milk now decorated the dash.

The driver ran over, looking like the twat that he was. 'Sorry,' he gasped. 'I dunno what happened there!'

He was lucky. Forgetting to put the handbrake on can kill you. A fellow driver, working for one of our competitors, was killed at the Barbican, in London, when he was crushed by his own trailer. We've all done it at least once, and mostly get away with it, just learning a valuable lesson.

I phoned Crouchy and got him out of bed. He listened as I explained the damage to him.

'So you can still make the tip?'

'No I bloody can't! The windscreen's gone and it's raining!'

Another tractor unit, an MAN, arrived at the truck-stop about 8am. Followed by a wrecker, an hour later. I carried on with the tip, and the DAF was towed to Glasgow for repairs. She was back on the road the following week.

Then, about a month later, I was running into Manchester late one night in a hurry, chasing a time slot and pushing 55mph. As I dropped down into an underpass, with concrete walls both sides and traffic alongside me, I crossed paths with a moron.

The concrete block he threw off the bridge narrowly missed the screen, but ripped off the air deflector before smashing into the bumper, pushing it right back into the passenger step. As the truck bounced right over the block, I managed to keep her in a straight line and bring her to a halt. To say I was a bit shaken would be underplaying it. If that block had come through the windscreen I would not be telling you this tale, believe me!

The police duly arrived and informed me that a double-decker bus had been hit just days previously.

'Did you see them?' asked the sergeant.

'No,' I replied.

'Were they black?' asked his colleague.

'I said I didn't see them,'

'Bet they were,' said the sergeant.

I got an incident number and ran to Trafford Park lorry park. Sod the tip. This was the second time this had happened to me. Remember Billy's wagon at the Dartford Tunnel? I do!

In recent years drivers have died from such attacks. Is it murder or manslaughter? For the unfortunate driver it doesn't really matter, does it? Even today my bum twitches when I see people standing on motor-way bridges.

I was shocked by the blatant racism of the police, but at the time I had other things on my mind. It was only later when I was recalling my woes to the boys back at Silver Road that it struck me. We all like to think our boys in blue are the best in the world. I always told my kids, if you're lost or in trouble then find a policeman. But when I got blown over I really felt the police involved were cold and uncaring. You'd think there would be some sort of connection between professional truckers and law enforcement officers, but there isn't. They treated me more like a suspect than a victim.

I could tell you loads of stories about the corrupt police in France, Spain and Italy, where it's common place to bribe one's way out of a bit of bother. But I've met several French truckers and a Spanish one, who have told me similar stories about our custodians of the law. A few rotten apples? Maybe, who knows?

Sometime later I was getting out a show in Southend. The *Barron Knights Summer Show*. The doorman was watching me back down a narrow alley alongside the theatre when I felt a jolt. Getting out I was horrified to find that the corner of the trailer had ripped open the boot of a new Mercedes car. The hole was so big I could see inside. The doorman denied any respon-sibility and quickly disappeared. Only to be replaced by the owner of the Merc, Pete 'Peanut' Langford, the singer from the Barron Knights! He was not very pleased. Nor was Gordon!

My bad luck continued with the 2700 when I returned *Postman Pat*, after a short tour, to the pro-ducer's house in Gloucester. It was quite a small set and was to be stored in his barn. Unfortunately it was snowing and my tractor wheels got bogged down as I manoeuvred around the barn. Being late in the evening, it was decided to wait till the next morning, then get a neighbour's tractor to pull me clear. Then he'd hang around in case I needed him to get me out of there, once I'd off loaded. Crouchy wasn't too pleased, but it was midweek, so not a total disaster. If fact I might even get a lie-in.

At seven the next morning there was a knock on the door. Stood in the snow outside was Pam Ayres! What?

'Would you like some breakfast?' she asked.

Would I? You bet!

She waited until I'd dressed, then I followed her through the snow, into the house. It transpired that she was married to the producer. I enjoyed eggs on toast, sat with her two charming and very polite children. She was very easy to talk to, and very sympathetic to my dilemma.

It's so nice when you meet these 'celebs' and find them as genuine as they come across on the telly.

Bonnie Langford was another. While touring *Peter Pan*, she was loved and admired by all the crew and cast. At every get-out she'd come down and thank the local crew, bringing a case of fizzy drinks and a box of crisps to help with the long get-out.

When the show was in Brighton, Jo and the kids came to see it and met Bonnie backstage afterwards. The kids loved her and she had a great chat with them as she signed their programmes.

Then when you tour people like Wayne Sleep or the Krankies, it's hard to find anyone that says anything nice about them.

I met up with Lenny Henry under completely different circumstances. I was sat outside the Theatre Royal in Nottingham, ready for a Saturday night get-out. Lenny Henry came out of the stage door along with a lady companion, and they climbed into the back of a waiting taxi, right in front of my cab. As the car pulled away, a drunk staggered past and nearly collided with the bonnet. The taxi driver hooted as he pulled around him. The drunk responded by punching out the rear window of the taxi. He ran off as the car stopped and Lenny and his companion scrambled out, covered in shattered glass. In their hair and down their necks. He was angry of course, but still threw in a few laughs.

One of the most amazing things I've ever seen on my travels was a rainbow in the middle of the night. There were three of us, running to Newcastle in the early hours. We pulled over for a break in a lay-by on the A1, just north of Wetherby, at about 3am. There in the night sky was this amazingly bright full rainbow. A full moon, sat low in the sky behind us, must have had something to do with it. But I've never seen or heard about anything like it before or since.

12

Check Mate and End Game

IN early 1991, the trip of a lifetime loomed. *Chess* was going to Norway, with me as the lead driver. The get-out, on the Saturday night, was from the Apollo Theatre in Oxford. The get-in at the Olavshallen Concert Hall in Trondheim was on the following Wednesday.

We studied the route for weeks. It takes a lot of planning to get the timing right. We'd run to the ferry at Harwich straight from the theatre. Then over to Hamburg. Drive up through northern Germany and Denmark, then another ferry from Hirtshals to Oslo. Then we just had to drive the 350 miles to Trondheim, Norway's ancient capital.

My fellow jockeys were Russ Russell, in another 2700, with Paul Prior and Neil Steggels in 2100s. Paul had my old beloved OGO. All of us had stowed lots of winter clothing, it was summer in Norway, but not as we know it.

The get-out went fairly well and we shut the doors on the last trailer at 8am. In all, 30 tonnes of equipment had been hand-balled onto our four 40' trailers.

We pulled out with my truck on the front door. When we hit the long climb at Watlington, on the M40, the 2100s flew by me and Russell, disappearing into the distance. The CBs rang out with ribbing and queries about the pedigree of the 2700s.

We stopped at South Mimms, but the truckstop was

shut so we had to use the main services. Shit food for shit prices. We had been warned that most things in Norway were expensive, everything else was bloody expensive!

On the weighbridge at Harwich it transpired that the 2700s were each pulling nearly seven tons more than the 2100s. The ribbing stopped.

The MV Hamburg was more of a cruise-liner than a ferry. With discos and bingo to help while away the 21 hours it took to cover the 600 miles to its home city. We sailed promptly at 6.30pm. Our meals were covered by vouchers and we had a quick shower before calling it a day.

Another voucher covered breakfast and by the time we docked at 1pm we were raring to go. But the doors on the ferry stuck and it took some 35 minutes to sort it out.

We ran out of Hamburg on the A7 north. On reaching Flensburg, on the German–Danish border, a helpful English driver helped us fill in the relevant forms and 30 minutes later we were clear and running up the E3.

At 2am, as the fog came down, we pulled into Hirtshals. We'd done 520 Ks since Hamburg.

The Colour Line ferry, *Jupiter*, sailed at 10am and covered the 150 miles to Oslo in nine hours. The cabins had a speaker in them announcing the bingo in four different languages every ten minutes or so. There was no way of turning them off, so sleeping was impossible. Our complaints were ignored.

The longest part of the journey on the ship was the slow sail up the Oslof Jorden fjord. We had land on both sides for at least four hours before we docked. There was another ferry on our front door, and a tanker behind us. We sailed along in convoy past very impressive scenery.

Oslo is further north than Yakutat in Alaska, but with the sun hitting us at 29°C, it was very hard to believe.

Visiting trucks pay a kilometre tax, but they took our word for our speedo readings. Everyone we dealt with spoke excellent English and all were very polite.

At 8pm I led us out of the harbour and immediately took a wrong turn. After a tour of the docks, Russell eventually led us out and through Oslo. The first few Ks had more ruts and potholes than a second-class truckstop. We bounced through tunnels and over rivers, and the 2700's ECAS air suspension came into its own. The air-ride seats, once adjusted to your weight, were a delight. There was a 70kph limit for all trucks and you ran with your headlights on.

We had until 6pm the following day to travel the 350 Ks to our destination, so we weren't going to push it. Take our time, enjoy the scenery, this was new territory for all of us.

The Norwegian flag flies everywhere, there seems to be at least one on every house. The country is half as big again as the UK, but has a population of less than five million. The streets seemed clean, not rubbish everywhere like you see around London.

The motorway ran out 30 miles north of Oslo. We stopped for our break at Dal, only just squeezing into the small service area. We took tea sat out at a wooden table, overlooking a large lake.

We ran another hour or so before calling it a day at a lay-by outside Lillehammer, the site of the 1993 Winter Olympics. Lillehammer that is, not the lay-by.

By 8.30 the next morning the sun was so bright we had our shades on and our roof hatches open. It had rained here non-stop for six weeks until the day before we'd arrived.

115

The E6 took us through the Ormtjern Kampen National Park. More than a century ago, these mountains inspired Edvard Grieg, Norway's greatest composer, to write the music for *Peer Gynt*. A nearby mountain bears that name.

At Dombus, the E6 takes a 90-degree turn around the foot of the Rondane Mountain. At 2100 metres, this was the highest so far and covered in snow. The trailers only just cleared the central crash barrier on the turn, and our wing mirrors were just inches from the rock face. It gets even tighter as we climb up into the Dovrefjell National Park. The trees disappeared as we got higher. The lighter 2100s pulled away and were waiting for us at the top.

Wow! What views! We were surrounded by much higher, snow-capped mountains, but we could see for miles. A solitary buzzard wheeled in the sky as the sun beat down on us.

The E6 took us straight up the main street of Trondheim and to the Olavshallen Concert Hall, where there was a Norwegian TV film crew waiting for us. They filmed us pulling in, then got us to run around the block a few times for more footage.

Everyone we met spoke perfect English – better than us most times. Apparently most Norwegians grew up watching American and British TV programmes. They all came with subtitles and taught the whole country how to speak English. It's the same in other Scandinavian countries. People loved engaging you in conversation just so they could use their English, as it was rarely spoken otherwise.

The get-in began. A huge lift, capable of holding half a trailer load, took the gear down a level to the stage floor. The fit up crew had flown in the previous day from England and directed what they wanted

116

and when. The local crew were excellent and very professional.

The get-in wound down at 10pm. Two trailers tipped, the others would wait until tomorrow. We took the trucks off to the local Havna Service Station, our home for the next ten days. Yes that's right, ten days off!

We shared the lorry park with a lot of Nor-Cargo wagons. One of Norway's biggest carriers, they were based at the services. Their domestic wagon and drags ran at 47 tonnes, while their artics did an eight-day return trip to the UK. All trailers had spread axles with rear steer, and snow chains are compulsory.

We handed in our Shell fuel cards, this gave us 24-hour access to toilets, showers, a TV and a kitchen with a microwave.

We sat on the dock edge at midnight and watched the sun going down. It was summer and the sun never went below the horizon. The land of the midnight sun. It never really got dark the whole time we were there. At two or three in the morning it was like dusk, then it brightened up again.

We had all brought winter clothes, only to find the country in the grip of an unprecedented heatwave. Unloading the last two trailers the following morning was almost unbearable as the temperature hit 30°C before 11am. Finishing by midday, we showered and ate at the theatre before heading back to the services.

The price of beer there, and all over Scandinavia, was bloody ridiculous: two to three times the price back in the UK. And food! It was a good job we were warned, we'd all brought tins and bottles in our supplies.

We were invited to the first night party at the show. We ate as much as we could, and certainly enjoyed the free drinks. We staggered back to the trucks in broad daylight at 4am. Seagulls, nesting on top of lamp posts, dive-bombed us. It was hard not to flinch and run.

We had ten days to kill. Not too hard. Thanks to Sten, our man at the theatre, we were given free access to a local hotel's pool, gym and sauna. It was like being on holiday, although we had to adjust to sleeping in daylight. By 8am, the sun turned the cabs into ovens, so there was little chance of a lie-in.

We set up our cookers in one of the empty trailers and began working our way through our supplies. It didn't bother me too much that they were all car-nivorous. But it seemed to bother them. First of all apologising, then telling me what I was missing. But just light-hearted banter, mind you. Eating burnt flesh gets some people off. Most of what came out of their tins was processed waste products. Cow's anus and bullock's foreskin! Yummy!

We took the units out a couple of times, just to keep them trim for the return hit. We'd have picnics on the edge of fjords, and climb mountains overlooking the city.

While running about we picked up skip on the CBs, all the way from England. Clear as day, we listened to a breaker from Bassett, near Southampton. We tried long and hard to get back to him, but the skip was only going one way.

Truck drivers over there are held in high regard. It can cost you more than £4,000 to get your licence and you have to be competent in mechanics and accountancy as well.

Speeding is definitely frowned upon in all parts of the driving community. Up to 10 Ks over the top and it's a big fine. Any more and you get an instant ban for a minimum of three months. More than half a pint and you're over the limit. You get a long ban, after you've served the statutory 21 days in a prison cell!

We all went to see the show midweek. The audience went mad, they loved it! Encore after encore.

We designed our own '*Chess* Tour' T-shirt. Unique and stylistic. We had one each made up at a local printers. A few days later we saw some of the crew wearing our design! The crafty shop owner had run off a few extra and most of the crew, English and Norwegian, had one!

We'd been out for 16 days by the time the curtain fell for the last time in Trondheim. As the punters streamed out at 11pm, we sat with the trucks at the backdoor, waiting to do our bit. Our next destination was Paignton, in Devon, although if we were lucky we could get a night at home on the way.

After such a long lay-off and the last night party, our bodies creaked and groaned as we tried to keep up with the stage crew. They were good and quick, and by 6am we were shutting the last trailer door. A record for the show.

After a shower and a succession of shaking hands with the crew, we set off down the E6 for Oslo, where we would be catching the ferry home.

We only managed three hours before our eyes told us to stop. We had the time, no need to push it. We parked up by a waterfall in the Dovrefjell National Park.

The sun woke us up at 8am. So bloody hot, we all had our doors open most of the night. You couldn't do that back home.

Running into Oslo the roads deteriorated noticeably. Again, on the CBs, we listened into the skip from the south coast of England. We tried, but couldn't get back to them.

At the Fred Olsen dock in Oslo, we paid up our

kilometre tax. We'd done 7000 Ks between us, leaving us with a bill for £670. They took our word for the figures.

The 10,000-ton ferry, the *Balduin*, didn't leave until 4pm the next day, so we had a 'Farewell Norway' meal that night in a plush restaurant. The bill of £176 included six bottles of wine. After a well-deserved lie-in, we boarded the ferry just in time for lunch. A lift in the bowels of the ship took us up to the top open deck. Apart from one other, we were the only accompanied trucks on board. The rest of the ferry was full of unaccompanied trailers and containers.

The cabins were two-berth, then you just had a mess room and a kitchen. With Spanish stewards and Norwegian deck crew, everybody ate together in the mess room.

That was when we discovered that the *Balduin* was a dry ship! Shock horror. No beer, no wine, not even a docking bottle was to be had. All Fred Olsen ships are alcohol-free zones. Pepsi or Fanta, that was the choice. We spent two dry days slowly plodding across the Atlantic.

Despite the chef smoking over the stove, the food wasn't bad. Self-service and as much as you could eat. Endless tea and coffee, and you could rummage through the fridge for a snack any time you liked. But no bloody beer!

There were a dozen or so recent videos to watch, but we spent most of our time on the bridge, where the friendly skipper put us through a crash course in navigation.

The *Balduin* docked in Felixstowe at 10am on the Friday, after a smooth 800 nautical miles. But it took us two hours to disembark, they took all the unaccompanied freight off first. Very frustrating as we all wanted to get home for the night before Paignton on the Saturday. Then we couldn't leave the dock for another hour as a computer wouldn't recognise our release codes at the gate. A few calls from Crouchy to Olsen eventually put it right. Felixstowe was the biggest delay of the whole trip.

Then the M25 screwed up at Godstone and we lost another hour. But we all managed to spend a night at home in our own beds before the weekend get-in at Paignton.

I kept a diary of the trip and later wrote it up. I submitted it to *Truck* magazine and they published it over two issues as a 'Long Distance Diary'.

I was elated, my first published article! Urged on by this success, I wrote up the tale of me being blown over on the M1. I sent it off to *Trucking International*, and they printed it under the banner of 'My Mistake'. My mistake being, I should have ignored the pressure from Crouchy and parked up.

I began writing more pieces, usually about my experiences and observations. I even started writing a musical.

At the time, I was touring the *Rocky Horror Show*. What a show! Outrageous and shocking, it's like a rock show with audience participation. Scantily dressed fans, stockings and suspenders no less, scream abuse and chant out all the lines. Every night was like a party. It was a 32-week tour and I got to watch the show most Saturday nights.

It was a one-driver show. I loaded and ran two step frame twin axle 45' trailers. The get-ins were mostly Monday morning affairs, so I usually had the time to run both trailers to the next venue myself. If not, like running Plymouth to Newcastle, I'd run one part-way for someone else to carry it on, then I'd run back and fetch the other. If I was lucky I'd get to kip for an hour or two before the get-in on the Monday morning.

I always carried the stage manager with me, Roddy Steele. He ran the get-out then slept on my bed as I punched out the miles to the next venue.

It was while I was rushing about amid all the mayhem of the show that I started to write *Manson: The Musical*. It told the life story of Charles Manson, responsible for the murder of Sharon Tate and others in California in the 60s. Set to music from the Beatles' *White Album*, it had scenes of sexual activity and graphic violence.

I can only think that the company I was keeping, along with the show, led me to believe that it was a really good play. After months of writes and rewrites I submitted it to the Royal Court Theatre in London. The response I got back was mortifying. They commented that putting on such a play would have the audiences throwing up and fleeing the theatre. They didn't mince their words. They thought it was crap!

It knocked me back for a while, but after a few second opinions, I got the message. The world was not ready for *Manson: The Musical*.

It was on the *Rocky* tour that I became a granddad. Justine, our eldest, gave birth to Kayleigh. A beautiful tiny bundle of joy. I like to think I make a proud granddad. Jo was in her element, volunteering for endless babysitting duties.

During a busy period, Gordon took on an ex-army boy one Saturday morning, and had him running a rigid from Bath to Edinburgh the very same night. When he hadn't shown up at the theatre by the Monday, the cast of the show had to perform in jeans and T-shirts because he was carrying all the costumes. After a couple of days and a radio appeal, the truck was eventually found down a country lane in Nottinghamshire. The driver was out cold in bed, surrounded by empty beer cans. He was an alcoholic who'd been kicked out of the army just weeks before.

Gordon was furious. The yard's car park was being banked up at the time to stop the early morning noise upsetting the neighbours. Legend has it that Gordon personally drove the JCB that buried the errant driver's car under the 8' high bank of clay.

I got a 12-week tour for Rowan Atkinson. An amazingly funny man on stage, but off it, very straight and serious. I collected part of the set from his home in Oxford. On his drive, the 14'-high trailer threatened to pull down his telephone lines. So he stood on a chair, holding a long broom, lifting the cables as I eased underneath them. It could have been a scene from one of his sketches.

I once did a location shoot for *Alas Smith and Jones*. They just never stopped cracking jokes. In the canteen, all those sat at the tables surrounding the pair were crying with laughter. On set, the cameramen had to stifle their giggles as Mel kept up a constant stream of one-liners. The director kept telling him off.

You rarely get a tip at the end of a tour, although I did get a personalised tour jacket from *Rocky Horror*. But the best thank you I ever got was from touring a Continental Airways promotion show.

Part of the promotion involved Gordon having a dozen or so trailers painted up with adverts for the airline. Full 40' jobbies. There was one with the New York skyline, another with Florida beaches, flamingos and palm trees. All bright and garish, all running around the country promoting the airline.

The tour itself was just two trucks, doing two to three cabaret style shows a week, for travel agents and their guests.

Usually at ballrooms and conference centres, they were extravagant, wild affairs with sit-down meals followed by disco dancers, games and prizes. We drivers got to eat for free and then enjoyed the show before the get-out. With three different venues a week, we stayed with the show for the whole of the six-week run.

At the end of the tour, every member of the crew got a bonus. An airline ticket to the States! And we were members of the crew! We had a choice, New York or Miami. My fellow driver wasn't that keen on flying and I bought his ticket off him for £100.

So Jo and I got a trip out to Florida. We hired a car and toured around. Watching ospreys and alligators in the Everglades, frigate birds and pelicans along the Keys and sitting at Lake Okeechobee as vultures ate a dead fish just yards from the car.

I've always been into bird-watching. I got my first bird book for my eighth birthday. Remember the I-Spy books? If you start your kids off young, they will carry on loving the natural world all their lives. Believe me, mine do, and Kayleigh has followed on, we love our walks around the Wetlands and Wildfowl Centre at Arundel.

While in Florida, Jo and I finally tied the knot. We got married in Naples, on the west coast, after 16 years together. The kids were a bit surprised when we got back, but of course they were very pleased for us. The day after we returned, I was back up the road.

One of the longest solo runs I ever did (without a trailer) was from our base at Ashington, close to the south coast, to Inverness in Scotland, more than 600 miles. I had to pick up a trailer full of exhibition gear from a whiskey distillery. It took me two and a half days – a great trip right up through the Scottish

Highlands. I followed the A9 up the Spey Valley, alongside the Cairngorm Mountains. At Aviemore I took a slight diversion and visited the RSPB reserve at Loch Garton. It was down a tight little lane, and the truck only just fitted in the car park, but I couldn't come so close without seeing the ospreys, could I? Although in Florida they're very common, over here they are extremely rare. At the time they were the only breeding pair in the UK. I watched through binoculars from the hide. You could see the female sat on her nest just a few hundred yards away. Wonderful! A redstart flitted about just outside the hide and you could hear woodpeckers like machine guns all around.

I collected the trailer from Inverness and ran it back down to London. Just in time to do a get-out at Croydon, and run it back up to Edinburgh.

Things began to change at Lucking's. In a move that surprised everyone, Gordon sold the firm to McGregor Cory, part of the Ocean Trading Group, in a big money deal. Gordon stayed on to help run things, as did Crouchy. The new bosses put their own staff in and there was lots of tension and bickering.

Gordon eventually left after one almighty bust-up. Perhaps it was hard for him to relinquish control after all those years at the helm. Perhaps he objected to taking orders from those he saw as his incompetent inferiors?

Was he pushed or did he jump? Subsequent court cases confirmed he was unlawfully pushed, and he got another pay-off. He then set up another company, Gordon Lucking, and pinched a lot of the work off his old firm using his contacts. A few drivers even left to join him.

The office was now being run by McGregor Cory representatives. They knew transport logistics, but not the way the entertainment business ran. They demanded everyone ran straight! Giving us numbered tacho cards so we couldn't lose any. And we had to wear uniforms, jacket and trousers. Although Crouchy was still there, his influence was on the wane. We all knew he wasn't happy and wanted out.

The end came for me after I pulled my back while off-loading steel stage work in Charing Cross, London. I was bent over double in absolute agony. The crew had to put a ramp up to the trailer so I could hobble off. I shuffled off to the hospital, some 200 yards away, taking 20 painful minutes to get there.

I'd like to say Crouchy bent over backwards to get me home, but you know I'd be lying. By the time I'd been checked over and sent away with painkillers and a sick note for two weeks' bed rest, another driver had taken over my truck and gone, just leaving a van for me. There was no way I could even sit behind the wheel, let alone drive a couple of hours home. I ended up lying across the seats for the three hours it took Jo to drive up from Worthing and get me.

As you can imagine, I was pretty pissed off by the treatment I'd received. But it got worse. I had two weeks off sick. I can tell you, in the seven years I'd been at Lucking's, I'd never had time off sick. Never. The sickness scheme that McGregor Cory had brought in was obviously better than Gordon's one. Average weekly pay for the first few weeks.

But when I got my pay packet, it had been reduced to the basic government minimum. Less than half of what I'd expected.

Upon complaining, Crouchy pointed out the small print. 'Manager's discretion.' He had decided not to pay me.

'Why?' I asked.

'Don't like your attitude, son,' he replied.

'What's that got to do with me being sick?'

'Manager's discretion,' he said. 'Take it or leave it.'

I appealed to the Ocean Trading Group directly and they countered by saying I'd criticised them in a magazine article, so they backed Crouchy. Then McGregor Cory gave me a hard time for going over their heads to Ocean.

I just thought that after seven years of hard slog and loyalty, I deserved better than that. Gordon still held the property rights to the yard, and after a few more court cases, he kicked McGregor Cory out by refusing to renew their lease.

The firm moved out to Chiswick, just off the A4, and those drivers who didn't want to move with them got a pay-off. Mine was just £900 for seven years' service. And it was taxed.

Thank you and goodnight.

13

Bailey Boys

LIFE without Lucking's took some adjustment. I had my weekends back after seven years of get-outs. I could take the kids out, play with our grand-daughter and meet up with friends. My whole sleep pattern had to change. And Jo had to get used to me being around the house again.

There was not a lot of work about, and what there was paid peanuts. Unemployment was a driver's worst enemy and the bosses' best friend. Drivers were ten a penny.

In late 1992 I got a start for DR Bailey, pulling out of Ford Aerodrome, just outside Littlehampton. They ran up to 20 fridge motors over the water, Spain, Portugal and Italy. General haulage out, fruit and veg back.

Now I may well have given you the impression that I only ever worked for crooks and bad guys. Well, you have to understand the times – that is how the industry worked. There were a few firms that stuck to the rules, but they were the ones that went bust, undercut by those that didn't.

A lot of Bailey's drivers were ex-owner-drivers who'd gone bankrupt. They all told similar tales. They'd mortgaged their homes to finance the truck then lost the lot, home and truck, in the downturn. And a lot of them had broken marriages as well, truck widows can only take so much. A few slept in their cabs in the yard, others in a small caravan.

The money was crap, so most drivers made up for it by bringing back tobacco and booze and selling it on. 'Plussing' was another little earner for them. When fuelling up in Spain and France you'd pump in, say 200 litres, but sign for 400, then split the difference in cash with the pump attendant. You could get away with it because you ran with your tacho fuse out a lot of the time so your mileage couldn't be tracked. Bailey's knew it was going on of course, but with such poor wages they were still getting a good deal.

Your days off were at their discretion. When they were busy you'd be lucky to get just a couple of days every two or three weeks. When work was slack you could be off for four or five days at a time. Unpaid of course.

A few years previous to me starting, one of the directors of the firm, served time for attempting to bring in a load of cannabis on a truck. He served his sentence at Ford Open Prison, just a mile or so from his home. Legend had it that most nights he'd go over the wall and spend his time at home.

It soon became clear that the driver was the lowest of the low at the firm. You were always being reminded that they had a waiting list of drivers as long as your arm. If you didn't like the way things were run, well goodbye!

For my first trip I double-manned a Seddon Atkinson with a guy nicknamed 'Teapot' after his favourite brew. When he stopped driving he'd have the kettle on before the engine had died. A great guy who taught me so much. Not just border controls and plussing, but how to cope with an Eaton Twin Splitter. I say 'cope' because that's all I ever did. I hated that bloody gearbox! What was worse was guys saying how much they loved it! You only used the clutch to pull away and to stop. The rest of the time you knocked her into neutral, blipped the accelerator and, in my case, rammed her into the next gear. But if she didn't go in first time, you ended up revving and crunching until she did. A real nightmare for me.

That first trip was down to Sagunto, just north of Valencia, in southern Spain. We were running down empty to collect a load of oranges. Not many firms ran out empty, but with Bailey's it happened a lot. Maybe the low wages they paid meant they still made good money.

A ferry from Portsmouth took us to Ouistreham, Caen, in France. Sagunto was a dodgy one hit from here for a double manned team.

'Only 900 miles,' said Teapot. 'Easy!'

I loved the open roads, we just ate up the miles. And restaurants and truckstops showed me just what we were missing in the UK. Free parking and showers, whether you used the restaurant or not. Good food, cheap prices, although veggies seemed unheard of south of the English Channel. It's a good job I loved eggs.

France seemed to be just one long motorway. The autoroutes took you around towns and cities. Although mostly only two lanes, they always seemed to flow well with very few hold-ups.

After stopping at the BP truckstop just north of Bordeaux for a brew, we punched on down into Spain with me at the wheel.

The European borders were now open. No longer did you have to queue to check out of France, then queue again to check into Spain. You just slowed right down at the border and drove through. If the customs or the police wanted you, they'd just signal you to pull over. If not, you just drove on through and went on your way.

One of the first lessons I learnt was that flashing lights have a completely different meaning in Spain. At a T-junction, a Spanish truck flashed me, so I pulled out. Wrong! Whereas in the UK it means 'After you', in Spain it means 'Fuck off, I'm coming through!'

I also learnt that Spanish drivers can be quite aggressive. The truck overtook me, the driver wildly gesticulating and blasting his horn. Then he came in on me, trying to push me off the road with his trailer. Fortunately Teapot was expecting it and had me braking sharply in time to avoid him. But there was only a foot or two in it. Welcome to Spain!

There were no phones in the cab, so you just phoned in when you were tipped or loaded. This appealed to me. The freedom of the open road. You were in charge of your day. You chose your own routes, and stopped when and where you wanted to. Pre-satnav and tracking satellites, you really were on your own.

Pamplona was my first real mountain. Hairpin bends, one after the other, and sheer drops down the mountainside if you got it wrong. The flimsy crash barriers wouldn't stop a bicycle. Fortunately Teapot was now driving. He pointed out wrecked trucks at the bottom of the gorge. Once they went over the edge, they stayed there. The locals would 'salvage' anything worth having, but the wrecks just sat there as a grisly reminder that death is only a few feet away. Memorials to the fallen are placed every few miles.

Being in a right hooker made it easier for the driver as he could see how close he was to the edge. But for me in the passenger seat it was a terrifying experience. Trucks coming the other way passed just inches from the wing mirror, I constantly flinched and lent away from the door, much to Teapot's amusement.

Once we got to top, Teapot climbed into his bunk and I took over.

'Just follow for Zaragoza,' he said. 'Wake me up when we get there.' What confidence.

I felt good. This was what driving was all about. I followed the N121, mountains one side, plains the other and griffon vultures soaring effortlessly in the sky.

Passing Tafalla, the sky began to darken. As Teapot slept on, the rain began, then it turned to slush, then snow. Visibility was down to maybe 50 yards within ten minutes or so. This wasn't nice, I'd never driven in anything like this before. I could vaguely see a truck up front, but his tyre tracks just disappeared in the swirling snow. I struggled to see the roads edge. I tensed up. Should I wake Teapot? I was out of my comfort zone, the oncoming snow made my eyes funny.

I woke up my cab-mate. He agreed I'd done the right thing and we pulled off the road into a restaurant. Just running from the cab to the bar, we got blasted with the cold icy snow, soaking our jeans and stinging our faces. From the warmth and comfort of the bar we watched out the window as the storm worsened.

This was a dilemma for Teapot. He had to make the decision whether to continue or wait it out. The road outside was practically empty. When a truck did appear, it pulled off into the lorry park. After an hour or so, the winds eased off, and a few trucks drove slowly by. Teapot suddenly got up and said 'Let's go. If they can do it, so can we.'

With Teapot behind the wheel we slowly drove out of the storm and by the time we got to Zaragoza the sun was out and we had the windows down trying to cool off.

We made it to Sagunto, then waited for two days before we loaded.

'This is normal,' Teapot said. 'You'll have to get used to just waiting around in a bar.'

Well, OK then.

The downside to this was, you always pulled out in the evening, after you'd been up all day, kicking around, counting down the hours.

Next lesson. We'd pulled the tachos and the fuse at Zaragoza on the way down. What with the snow and my slow driving, we'd needed an extra couple of hours. So on the way home we pulled over, again in Zaragoza, and put in new ones. Now we could get back to the ferry at Ouistreham in one easy double-manned hit.

Teapot hit the bunk, telling me to stop at Pamplona and he'd take her down the mountain. Sounded good to me. Except by the time I got there it was snowing again. I woke him up. God it was bloody awful. Blinded by the oncoming snow, all I wanted was for Teapot to take over. But it wasn't possible. The road was deep in snow. Lay-bys were invisible, and with a truck tailgating us, there was nowhere for us to pull over safely.

Before I knew it I was on the 9 per cent descent. Teapot was as nervous as me, he started giving me advice. And believe me I was glad of it! We were travelling on packed snow. Braking was out of the question, if the wheels locked up we were in real trouble. Once I was in a low gear I didn't want to chance it by changing up or down. The exhaust brake was the key, but too much and she would stall. Then you'd lose your power steering and the contents of your bowels!

Teapot called out when to hit the exhaust button on the floor. 'On!' he'd shout, then 'Off! His sweating brow didn't help my confidence, or the fact he'd put his seatbelt on. Hairpin after hairpin, and the descent went on. A truck in front of us gave me a guide through the blinding snow. His brake lights hardly came on at all, he was just exhausting it as well. Teapot explained that a lot of Spanish trucks had retarders on their trailers, holding them back, rather than braking from the front.

It took me over an hour to get down below the snow belt, possibly the longest hour of my life. I knew we were back on safe terrain when Teapot put the kettle on and made a brew. He congratulated me, and I thanked him profusely for his help.

Stopping at the border, we fuelled up. Teapot plussed, then used the cash to stock up with tobacco and beer, filling the bottom bunk.

After another big hit to Caen, we caught the ferry back to Portsmouth. A six-hour crossing, long enough for a quick shower, a free meal and then a welcome few hours' kip.

Arriving at Portsmouth, I fully expected to go back to the yard then home. Wrong. A truck and trailer was waiting for me in the docks. A well-worn Seddon Atkinson. I transferred my gear over and was on the same ferry back out, on my way down to Sagunto again.

Although I was a bit nervous about my newfound responsibilities, I relished the challenge. I felt like a proper continental driver at last. It felt good climbing Pamplona on my own. No snow, thank god, but I did have a problem soon afterwards. A puncture on the trailer just outside Zaragoza.

Now I know I'd changed wheels before with Billy, but this was different. Sat on the side of a busy road in the soaring heat, I struggled with the huge super single wheel. My weedy ten stone jumping up and down on the bar, had little effect. It took me well over an hour of sweat and toil to make the change. In that time, several Brit drivers passed by, each beeping their horn and waving. I'd heard all about Brits abroad looking after each other, but on that day it seemed just a bloody myth.

This truck also had the dreaded Eaton Twin Split! I was really shit at coping with it. I'd let her run as long as I could before attempting to change. Now the speed limiter, when connected, came in at 90 Ks. And when it did, the accelerator pedal went flat to the floor, only coming back up again when your speed dropped down.

So there I was, enjoying the scenery, listening to Bob Dylan on the sound system, cruising down a long slow decline just south of Teruel. She was really moving, fully freighted and keen to get home. As she crept up to 90 Ks I decided a half gear down change was necessary, just to hold her back a bit. I knocked her into neutral, went to blip the accelerator ... but it wasn't there! The limiter had cut in and the pedal was flat to the floor! Not being able to blip the pedal meant I had even less of a chance of engaging a gear! I was flying down the hill in neutral at 90 Ks! I was crapping myself as I stood on the brakes. And I mean stood on! I was up out of my seat, literally standing on the brake pedal. She shook, she shuddered, I was screaming out to her 'Come on! Come on!'

I was really considering my options, stay with her and leave the road at the next bend, or get out the door now! My panic-stricken braking eventually slowed her just enough for the limiter to disengage and the accelerator pedal popped back up. Knowing I'd only have one stab at it, I revved like mad and slammed her in. By some miracle she went in and the rev counter nearly shot off the clock as she backed off. In the mirror I could see smoke coming from the brake drums, but I didn't care, I'd got away with it. The first lay-by saw me pulling over to let my jelly-like legs recover. That was close, too bloody close!

There was a lot of truck crime in Italy and Spain. Thieves knew that foreign truckers always carried a lot of cash for péages and other expenses. You heard so many stories it was easy to become paranoid about it.

Not long after I'd started I was in Mercia, collecting onions the following day. I parked up for the night opposite a restaurant. While enjoying my frittata and watching some footie on the TV, some lowlife broke into the truck. Because I carried everything important – passport, money and licences – with me in a bum bag, they 'only' got away with my camera. But in doing so they smashed both side windows.

Incredibly Bailey's wanted me to run home and sort it out back at the yard.

'What about the windows?' I asked. 'It's raining!'

'Stick some cardboard in,' replied the boss.

'So how will I see in the mirrors?' I replied.

I ran into a DAF dealer in Madrid and got new windows fitted the following day.

Soon after I changed motors for a bright yellow MAN. And it had a proper gearbox. How I loved that truck. A driver and his truck, eh? They depend on each other for their very survival. On long runs I'd often talk to her. Encouraging her up steep mountain-sides, and even discussing routes with her. Some might say I was 'cab happy', others just assumed I was weird. What would Freud make of it, I wonder?

A big difference between UK running and over the water is the way truckers are perceived. On the continent, truck driving is rightly seen as a top professional job, and drivers are shown respect and gratitude.

My first convoy. That's me bringing up the rear aged 4.

DAD — Norway
xmas 45

My father in Norway, 1945.

MV *Rauhine* off Jamaica, 1967.

SS *Amberton* off Thailand, 1968.

'No fucking chance.'

Happy families in Knighton: Justine, Daniel, and baby
Dylan.

SPS: great cab, shit gearbox.

Sussex Haulage DAF 2800, pre-jackknife.

Wing mirrors in! Pebble Mill film location.

Queen Vic, *EastEnders* set.

Propping up the government.

Sad end for
Haymarket.

Old Vic. Better known as OGO.

It's a Knock Out!

Chess get–out at Oxford.

Chess get-in at Trondheim.

Trailer cuisine: Neil, Russell, and Pete.

Continental Airline tour.

Continental Airline tour.

Wedding day in Florida.

My new motor!

BBC filming unit, Edinburgh.

Helping the war effort.

Bloody
phones! Well
done son.

Wide load at the Barbican.

DR Bailey DAF 2800.

DR Bailey Portuguese A-road.

Murfitt's, Lake Como.

Descending the Blanc.

Stuck up to the axles in Portugal.

All together now, pull!

Jo with our Eurostar.

Foggy's DAF 95.

Kayleigh.

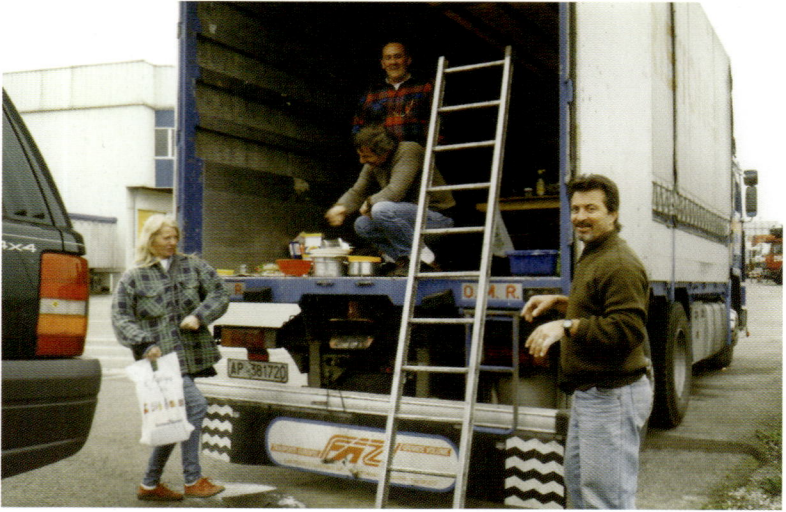

Jo with fellow detainees in Santander.

The blizzard lasted three days.

Clearing the snow off before we could pass through the Blanc.

Jo takes us through the Blanc Tunnel.

These truckstops get worse!

Taking a break: happy times.

Europa's FH12: great truck, great job.

Jo cleans the back lights.

Kayleigh. Getting bigger every day.

VW trip to Sweden: Jo, Kayleigh, and little Patrick.

Agency work for DHL out of Southampton.

The beginning: Daytona Beach and the Atlantic.

The end: Dillon Beach, San Francisco and the Pacific.

Harvey the
RV.

Sequoiadendron giganteum: a bloody big tree.

Snowboarding in Breckenridge.

Now that's what you call an RV!

Father and son at the Grand Canyon.

Wilderness.

New Scania
with Maritime.

Sandy Bruce
at Ford,
Littlehampton.

Talisker.

Charley and Dan aboard *Talisker*.

On the Cut.

Cold winter outside. Snug and cosy inside.

Me and Che.

Cuban taxi. Give us a go!

Trannie on tour: Robin Hood's Bay, Whitby.

A close call!

Still smiling after all these years.

In both France and Spain, kids still grow up wanting to be truck drivers. 'Camioneros', sounds much better than 'lorry drivers', don't you think?

In most garages in Spain, when a truck pulls in for fuel, an attendant will do the business for you. And then he'll wash the flies off your windscreen, fill up your washer bottle and even check your oil, if you want. And you're often given a free bottle of spirits or a king-size bar of chocolate. They want you to come back, and they run very good loyalty schemes. No way would they dream of charging you to park up for the night on their forecourt.

The Spanish are very religious, they truly believe that the almighty will protect them from all evil. This shapes their attitude to driving. On the side of their trucks they often have images of the Lord, and along with crucifixes and rosaries hanging in the windscreen, these tend to substitute for insurance and driving skills.

They can overtake on blind bends because the Lord tells them nothing is coming the other way. And if it does, then he will provide the escape route. Then, if it all goes wrong – and believe me it often does – they'll be able to sit around in that great truckstop in the sky, with free parking and showers, blaming someone else for eternity.

My reversing skills had improved some over the years, and my confidence in my abilities had grown. Then one day I had another nightmare come true.

You'll recall my horrific experience with Brisbane's at the treatment plant, well I nearly shit out again in a small town in Spain, close to the Portuguese border. I was looking for my tip and had stopped to ask directions. I thought I was so fortunate when the woman I asked spoke perfect English.

Just follow the road into town, she said, and at the T-junction turn left. The warehouse was down there on the left. Great! Satnav was sci-fi in those days. Maps would take you to the area you wanted, but you relied on the locals to guide you in.

I'd started off the day full of the joys of spring. Eagles hung in the sky and oleanders blossomed along the roadside. The temperature was in the high 80s and I had my T-shirt and shades on. Very cool.

I drove slowly down the narrow main street into the town. Cars parked one side, shop canopies just inches from the trailer on the other. A slight bend had my heart racing and required a shunt to bring the trailer around a parked car. Following traffic had to back off. Rounding another bend I was relieved to see the T-junction ahead. Then my heart sank to my boots as I realised there was no way I could possibly turn left or right. It was too bloody tight! The road was more like a bloody alleyway, even a car would have struggled to make that turn.

The blood drained from me, and my bowels were threatening to complicate matters even more. I stopped and got out on shaky legs, trying to assess the situation. Cars behind me began tooting their displeasure and a small crowd gathered. People began offering advice and although it was all in Spanish, I got the message. I was going to have to reverse out the way I'd came in! I wanted to die! Shrivel up and be transported to my bed at home where I was sure I'd wake up and find it was just a bloody nightmare.

I recalled the tale of a poor Lucking's driver who shit out big time in a tiny village in Cornwall. They had to get a crane in to lift up his wagon and turn it around. My, how we took the piss! If only that crane was here, now.

A policeman on his motorbike arrived and I

thought, it's even going to cost me money now. But no, even though most Guardia Civil would fine you £50 just for a bulb out, this guy was a saint. He laughed and joked as he pointed out, in his pigeon English, the rear of the warehouse just 100 yards away. Then he too started pointing behind me.

I watched as he cleared the traffic away from behind the truck. Some of them struggled to reverse back up the street, but it gave me no comfort at all. Surrounded by a large group of kids, enjoying the spectacle, I remounted and slowly reversed up to the first bend. It was blind-side, so I had to rely completely on my friendly policeman. I took four shunts to round that bend. Reversing back until he blew his whistle, then shunting forward as far as possible before going back again. Shopkeepers had to pull in their canopies, locals came out to move their cars, for everyone except me it seemed like carnival time. Nobody seemed uptight or even mildly cross, but I was stressed right out. And it was so bloody hot. The sweat was running down my nose and into my eyes.

After a short straight I began negotiating the next, and final bend. This took even more shunts and now I had kids holding on to the wing mirrors and taking a ride. They were laughing and having a good time.

It took me well over an hour to reverse clear of that dreadful place. I then followed the policeman on his bike as we circumnavigated the town and came in from the other side.

I don't believe in angels, but if I did, then that man was my guardian angel. I couldn't thank him enough. I shook his hand until he had to prize me off. It's a good job he didn't ask to see my tacho!

14

Jo Hits the Road

JO began to accompany me on some of my trips. Bailey's allowed you to take kith and kin with you as long as it didn't interfere with the job. The kids were growing up fast. Justine and Kayleigh had their own place now, Daniel was an apprentice carpenter and Dylan was in his last year at school. Jo had done her bit, now she was on time off. The kids all agreed she deserved it and so did I.

Because even Bailey's drivers couldn't drive on Sundays in Spain, France or Italy, we'd usually get a 24-hour break at least once a trip. I tried to make sure we took these breaks out of town, in the countryside and small villages. We both loved our birdwatching and had our own names for our favourite stopovers.

There was 'Vulture Valley', just north of Madrid: a restaurant on top of a hill, overlooking a steep valley where you could actually look down on griffon vultures soaring on the thermals. Then there was 'Golden Eagle Mountain', near Granada, where you were guaranteed to see these beautiful birds following the high ridges. The first place we ever saw golden orioles was at a village near Barcelona, thereafter called 'Oriole Pass'.

We started doing Italian runs, nectarines and peaches back to Covent Garden via Calais. Running over the Alps, through Fréjus and the Mont Blanc tunnels, we saw Alpine swifts and ravens flying high over such incredible mountain scenery.

Jo loved it so much that on her 40th birthday I bought her HGV driving lessons. In those days you could go straight from a car licence to a Class 1, and that's just what she did. Jo struggled a bit with her reversing, just like I had, and took a couple of attempts to get it right. But in the end she got it and Bailey's then employed us as a double-manned team.

It was a job we both loved and a turning point in my career.

But Bailey's certainly didn't do us any favours. They only paid Jo night out money, about £12 for every night away, and then only when we were over the water. In the UK she wasn't paid at all. When you realise that we were doing 22 hour days, if we ran legal, you can see the benefits were all one-sided. But with Jo as an inexperienced driver, and with no other work about, we went along with it. We enjoyed the positives and tried to live with the negatives. It wasn't always easy. Let me give you an example of man-management, Bailey's style.

One day I'd gone into the office on our day off to collect our wages. A glass hatch in the drivers' room opened into the transport office. Sat just the other side of the hatch was the transport manager, Bigshit. Real name Bignall, but always referred to by his nickname. He was sat at his desk, newspaper spread out and lunchbox open.

I slid open the hatch. 'Morning boss,' I said cheerfully. 'I'm here for our wages.'

'Can't you see I'm having my lunch?' he snapped, slamming shut the hatch.

I watched through the window as he ate his sandwiches. When he'd finished reading the paper he began to do the crossword. After about 15 minutes of me waiting, he folded up the newspaper, slowly and deliberately, then put away his lunchbox. Then reaching into his desk draw, he took out our wage packets and, leaning across his desk, passed them through the hatch to me. He didn't even bother to look up.

At the risk of repeating myself, I will. In those days you put up with unbelievable shit and conditions. Take it or leave it. Job or dole? It was your choice, as you were forever reminded.

We now ran legal most of the time. Double-manned, there was no reason not to. We could do a 22-hour spread over, with up to 20 hours driving, have an eight-hour break, then do it all over again. It was hard work but we loved it.

Just like Lucking's, there was a heavy drinking culture among the drivers. On days off, or while waiting for a load, the common practice was to head for a bar and get legless. Bragging rights went to those who, after just a few hours' kip, still made the ferry on time.

Some of the drivers did themselves no favours at all by sucking up to the management. It was the old 'well done son' syndrome. On several occasions, usually Italian runs, we'd load and run non-stop back to the UK, only to be beaten back to the ferry by single-manned drivers who'd left at the same time as us. With king-sized egos, they'd boast and take the piss.

Of course Bigshit would slap them on the back, then turn them around and send them back out again. We saw these guys as pawns in a game, they were being used and didn't have the sense to see it. They never got any medals or extra pay. They just risked their lives, their licences and their livelihood for what? Brownie points?

And it wasn't only on the road that the Bailey Boys risked their lives. One tale tells of a driver, who after a long hard slog down to Torino, was so cold in the cab he left his gas cooker burning to keep him warm through the night. Unfortunately the flame blew out and the gas nearly killed him. He awoke feeling like shit and with a splitting headache. Sitting up in bed he light a fag. Boom! The blast blew him through the windscreen and out into the road. He survived with just a few scrapes and bruises, and an ear-bashing from Bigshit!

Apart from all the crap and aggravation from the office we enjoyed the job. One of our favourite runs was to Odemira, in the south-west of Portugal. Tiny back roads and running down dirt tracks really tested your driving skills. Suicidal drivers and potholes the size of mineshafts kept you on your toes.

We'd sit in the corner of a field as ancient machinery ploughed out potatoes and beetroot, which was then loaded onto our fridge trailer. A flock of bee-eaters followed the tractor up and down the field. Wonderful multi-coloured birds – red, yellow and green, like large humming birds. These birds are seen on rare occasions on the south coast of England, causing a frenzy among bird watchers.

Sometimes we'd be sat there for two or three days before we were full. We'd drop the trailer and run to the local bar in the tractor. We were there so often they treated us like family.

Then, usually late in the evening, we'd set off for home, one-hitting it to the ferry at Caen, nearly 1,200 miles away. This was one of the few runs where we'd have to lose a few hours. We'd run out to the Spanish border before slipping the fuse back in. We weren't that happy to do it, but that's the way it was.

Often we'd land at Portsmouth, swap trailers in the docks, and get back on the same ferry to go back down again. Thus managing to get a 'legal' break.

Days off at home were not very frequent. Then never more than two at a time, often pulling back again out in the late evening. We felt like nomads, forever on the move. Always going to or coming from somewhere. But we thought it was great! Cab happy? Maybe.

We'd been running as a team for nearly two years when a crisis at home stopped us in our tracks. Daniel, our eldest, was taken seriously ill and needed a kidney transplant. Jo donated one of hers – a truly selfless act from a loving mother to her son. The operation was a success and although Daniel recovered pretty quickly, Jo's part in it meant she was laid up for several months.

I'd like to say that Bailey's were sympathetic and understanding, but they weren't.

'Give us a call when you're ready to come back,' said Bigshit, 'and we'll see if we've got a truck for you.'

With the help of Jo's mum Kathleen, along with family and friends, I was able to return to work after a couple of weeks, as a solo driver. With no holiday pay to call on, our flimsy finances had nose-dived.

When I called Bigshit to say I was available again, not once did he ask about Jo or Daniel. He just gave me a truck reg, a trailer number and a ferry time.

I ran solo for a few months before Jo was well enough to return, although I still seemed to be running the same long hours that we had as a team. On one of the runs I made Bigshit a very happy man. Unintentionally of course.

Please believe me when I say that, basically, I'm an honest sort of guy. I taught my kids the difference between right and wrong, and I try to live by the same principles.

You can see I'm about to make a confession, can't you? Well …

Running for the ferry, late one night, up from Italy, I pulled off into a large Shell station just short of Calais. I was truly knackered and looking forward to some overdue time off. I manoeuvred in between two pumps so I could fill both 400 litre tanks at the same time.

Off I went into the garage with my fuel card. The young female attendant sat with her back to me at the counter. She was on the phone, gabbling away at ten to the dozen as only the French can do. Her conversation didn't miss a beat as she swung around on her swivel chair, snatched the card from my hand, then swung back away from me. No words of acknowledgement, not even a nod or a wink.

Now I've never been a Francophobe, despite all the grief we get from the waiters on French ferries. Or the French bar staff who ignore English accents in a crowded bar. It's nothing personal, it's just part of their culture to be rude and arrogant to foreigners. You just have to accept it and move on.

But yes, this girl did pee me off a bit. After a tough hit up from Milan, a smile or even a quick 'bonjour' would have been nice.

I went back out into the night and filled up both tanks, over 700 litres.

Back at the till, she was still at it. Now, much to my shame, I don't understand a word of French, but don't they have full stops or pauses? Whoever was on the other end of the line, didn't seem to get a word in. Perhaps she had another job on the side. Running a phone sex chat line, maybe. Her conversation never faltered as she swung around to face me. She swiped the card and pushed the receipt over for me to sign. I sign for one pump, 360 litres, then waited patiently for the second pump. It never came. She took the pen from my hand, then swung back away from me.

I paused for a moment or two as she continued to gabble on. Then I turned and walked out, calling from the door, 'Have a nice day!'

All the time I was expecting her to come running after me, to correct her mistake, and I would have willing signed for it. But she didn't. That could be the most expensive phone call she ever made.

Unfortunately, I was heading back to the yard, so I never gained personally from her rudeness (or my dishonesty, depending on your point of view) except maybe brownie points from Bigshit. And he never heard any more about it.

Jo's recovery was slow but sure. The operation was a big one, leaving a huge scar on her side. She'd tell people it was a shark bite. She eventually got back on the road and we carried on with our adventure. We ate a lot of cab cuisine. Tins of beans and rice pudding, when we were on the move, and soups and rolls on short eight-hour breaks. We had a 16-minute electric kettle, we'd plug it in 16 minutes before we were due to park up. Then it was handbrake on and pour the tea. I learnt that one from Teapot!

We always made sure we feasted at restaurants on longer breaks. Food was cheap and on the whole pretty good, if you like eggs and cheese.

We used the Eurotunnel for the first time and hated it. Endless queuing, trains cancelled and delayed: the ferries often seemed quicker and more reliable.

Occasionally we'd ship out of Portsmouth to Santander in northern Spain. A 36-hour ferry that gave you a bit of time to relax and catch up on your sleep. It also counted as your weekly break. Then we would hit Portugal from the north, running down the west coast. Great sea views but really crap roads. The potholes around Lisbon must be the deepest we'd ever seen.

We just loved Portugal – the country and the people. Days spent waiting to load were never wasted, we'd go for walks bird-watching, and sit in pavement cafés people-watching.

On one such trip we were running home and passed through Badajoz, on the border with Spain. It was maybe five or six in the morning and we were surprised to see crowds of people out on the streets. We passed this way often and this was most unusual. The traffic lights were out and as we cautiously turned to run out of the town we noticed a demolished building alongside the road. Then another. There was a lot of debris in the road, branches and rocks, and thick mud everywhere. People stood around seemed to be in state of shock. As we drove out of the town we saw upturned cars in the middle of fields, then a lorry container upside down on the verge of the road.

It transpired that due to heavy rain fall, the river that ran through town was swollen to capacity. A small bridge, just west of the town, became blocked with debris and fallen trees, causing a dam across the river. The water built up behind it until the force of the current smashed the bridge away and a tidal wave swept down into the town like a tsunami. Four houses were swept away and 11 people died, just a few hours before we passed through. Mother Nature takes no prisoners.

15

Story Time

IT was about this time I started writing my first novel, *Diesel Rose*. Handwritten on notepads on days off over the water, then typed out on a trusty word processor when I got back home.

Diesel Rose was a lady trucker. Big, mean and very aggressive. Our poor hero, Dave Swan, is used, abused and conned by her in every possible way. Sex and drugs and rock 'n' roll. It took me many years to finish and I eventually self-published with Authorsonline.co.uk. It got a great review in *Truck & Driver* and sold quite a few copies. Although in the hundreds, not thousands. It's a great feeling to see a book with your name on the cover.

At the same time I was getting articles regularly published in various truck magazines. Some were about my experiences on the road, others my rantings on rubbish in lay-bys and the state of so-called 'service stations' on motorways. But don't get me started!

On a cold winter's day in a lay-by, on the Spanish border, I was attacked by a wild animal! I was having a quick pee against the wheel, before setting off after a break, when I heard a faint mewing. Following the sad distress call, I tracked it down to my drive wheels. There, in-between the tyres, was a young, fluffy, black and white kitten. Ahh, I thought, reaching down to pull it out.

'Ouch!' The bloody thing gouged a couple of grooves out of my hand as it spat and hissed back at me.

Once I'd finished my cursing, I wiped the blood away down my jeans and wrapped a lot of bog roll around the scratches. I then realised I had a bit of a predicament. The aggressive feral kitten had been seeking the warmth of my tyres and had wedged itself up between the twin wheels. If I'd driven away, it would be dead, and my hand would not be throbbing. But I hadn't and it was, and now the kitten had to come out, didn't it? My leather gloves were only effective until it sank its teeth in. More blood, more cursing.

I had a tetanus jab 20 years ago, does that still work? I wondered, as I grabbed its tail and pulled. It squealed like hell, in Spanish I presumed. I spoke gently and pulled harder. It suddenly came flying out and I managed to let go before the snarling teeth found its next target. It scampered off with neither a please or thank you.

As I smeared antiseptic cream all over my wounds, I wondered why I'd bothered. Although we'd had cats while living in Wales, I wasn't really a cat fan. No bird-watcher is. A single cat can kill thousands of birds and mice every year – more so if it's feral. With an estimated 11 million cats in the UK, well, you do the sums. The domestic cat is an invasive species to our shores. Our native wildlife struggles to cope.

In France and Spain most lorry parks and lay-bys have colonies of feral cats that live on the waste and garbage we humans leave behind.

In the countryside of Spain you often see packs of feral dogs, mangy and disease-ridden, running wild and keeping the shepherds and goat herders on the lookout. Reverting to type, they'll rip apart anything edible that crosses their path. They're not too good at crossing roads though. The rotting carcasses of flattened dogs litter the highways as they slowly get devoured by vultures and maggots. Nothing is wasted in nature.

The vast Spanish landscape is just amazing. Mile upon mile of arid countryside, bordered by often snow-capped mountain peaks. Rivers carve canyons and valleys hundreds of miles long. The Americans filmed early western movies here. It's wonderful wilderness that makes you feel good just passing through it.

But unfortunately the locals often don't seem to appreciate this natural habitat. While their cities seem so spotless and clean, they tend to use their countryside as a rubbish tip. On the sides of roads, miles from the nearest habitation, you'll often find household waste such as fridges, TVs and old furniture. Building rubble, old cars and even food waste pile up and attract even more. If a car is wrecked in an accident, it's stripped of anything of value, then just left. Trucks that fall off mountain-sides are also left where they land, reminding us truckers what can happen if we don't stay focused.

It's always been difficult being a veggie on the road, even more so abroad. In Spain you're laughed at as you sit under hanging, curing hams at the bar, and the French openly take the piss. But it never really bothered us too much.

There is a lot of hypocrisy among meat-eaters in the UK. Recent stories in the press caused outrage when many people suddenly discovered that they'd been eating horse meat for years without knowing it. But I have to ask, what's the difference between a horse and a cow? You'll happily eat one, but squirm at the thought of eating the other. And so-called animal lovers campaign against people in China and the Philippines eating dogs. But why not? These people will willingly eat lambs and calves without a moment's thought. They'll enjoy a goose for Christmas but recoil in horror at the stories in the press of Eastern Europeans eating swans. It seems a very British thing, some animals are for eating and others for stroking. And never the twain shall meat!

These same people see us veggies as weirdos, just following a fad or craze. Maybe we just see the things that they don't want to.

My feelings are re-enforced by my own experiences, as recalled in Newport, South Wales. Remember the guy with the walking stick? Let me tell you about another incident that, for me, sums up the whole meat trade.

Transporting cattle and sheep across Europe must be one of the most awful things man does to these beasts. Because they're going to be slaughtered when they arrive at their destination, nobody seems to give a shit about their welfare. Cattle truck drivers can run much longer hours than the rest of us, and they do. Laws about watering the animals and giving them enough space and rest were never heeded because no one really cared. Cargo is just cargo, living or dead.

The laws in the UK were eventually changed to limit the number of animals a truck could legally carry. The meat industry was up in arms. 'Jobs will go!' screamed the farmers. 'Firms will go bankrupt!' yelled the hauliers. Of course none of this happened because a lot of hauliers found a way around it.

On a trip to France one evening our truck was loaded onto the ferry alongside two cattle transporters from Wales. An artic and a six-legger rigid, both full of bellowing cattle. The cries of the stressed out animals

151

used to really bug me. More so because nobody else seemed at all bothered.

On arrival at Caen we had to park up for an hour or so before we could legally run. The two cattle trucks pulled off the ferry and into the car park alongside us. They backed up close together, then the cattle from the rigid were trans-shipped over to the artic. Illegal in the UK, yes, but not in France or Spain. The empty rigid then drove back onto the ferry and went home. The overcrowded artic then set off for a long non-stop drive down to Spain.

The farmers will tell you that, over the years, animal welfare has changed. Yes it has, but only because changes in the law, brought about by activists, has forced them to. Most farmers fought the changes all the way. And there's a lot more to be done.

We humans learn from our mistakes, that's how we've evolved so far, so quickly. We've stopped sending children up chimneys, we no longer see other races as inferior and we're finally recognising the damage we've done to our planet with the industrial revolution. Now the exploitation of animals, fellow species on our planet, needs to be seen from a different angle, looking at welfare rather than greed.

We enjoyed our time at Bailey's despite all the attitude problems and lousy pay. Jo got her driving experience and I could now claim to be a continental driver.

The end came for us as we were thinking of moving on. Friends of ours were double-manning for international haulier Murfitt's, and they were getting paid a wage each! So when we got back from Portugal one Christmas Eve and were told we had to ship out to Italy on Boxing Day or else. We chose 'or else'. Thanks for the memories.

16

Muppets

MURFITT'S was a large international haulage company with its main depot in Peterborough. When we phoned for an interview, the transport manager suggested we bring our gear along, just in case. Would we be OK for an Italy tonight?

We got the feeling they were desperate. But we got a start and began running an Iveco Euro Tech on regular runs down to Italy, with the occasional run into Spain. The wagon was a bit underpowered and the cab a lot smaller than we were used to, but it seemed to be the perfect job. We were both being paid a good wage and, despite the rumours we heard, we were never put under any pressure to run bent. The firm was divided up into several divisions. Wagon and drags ran mainly to Scandinavia, mega bulk artics to Italy and a few tilts and taut-liners did Spain and the rest of Europe. Although everybody did seem to mix and match.

We were treated as human beings and felt the job was the best we'd had. They had five or six double-manned teams, mostly husband and wife, and they all seemed happy with their lot. Other drivers always referred to Murfitt's as 'Muppets'.

We actually ran out of Dover. We'd leave the truck in the docks, running home in our car from there, a good two-hour hit in the early hours. We rarely went back to Peterborough.

Murfitt's drivers had a reputation for wild behaviour. The stories we heard were of drunken weekend stopovers where the drivers would fight one another, and anyone else who fancied their chances. We never saw any evidence of this at all; maybe it was just a myth and a self-perpetuating legend. I think most of the guys and girls revelled in these stories. Also, despite the tales of wagon and draggers hating artic drivers and vice versa, we never had any bad vibes from them at all.

The tales of bent running did, however, seem to be true. Huge fines, paid up by Murfitt's, seemed a regular occurrence for the wagon and drags, and drivers running home on the following week's tachos seemed fairly common. Again the drivers involved used to boast about it, like it was an honour to work all those extra hours for nothing.

We enjoyed the Italian runs, weekending in Aosta, just south of the Blanc, was a favourite for us. Lots of eagles and hot air balloons, with spectacular mountain scenery all around. We even managed a day off parked up alongside Lake Como.

We'd often see the Italian Fire Brigade on the autostrada. Running on blue lights, and with sirens wailing, being overtaken by all and sundry as they raced around at a truly pitiful speed. You wouldn't want to be in a burning building waiting for them to arrive. A view that was re-enforced one day as we left Rome.

We heard the wailing from behind us, as the fire engine battled its way through the heavy traffic. When they got close I pulled over, half onto the pavement, just in front of a petrol station, to help them ease by. The fire engine drew level, then pulled across the front of us into the garage.

The fire must be in the garage, we thought,

though we could see no smoke or sign of distressed people.

A fireman got out of the engine, lights still flashing, and walked across the forecourt into the shop. When he emerged several moments later, he was unwrapping a pack of cigarettes. After dropping the packaging on the ground, he stuck a fag in his mouth, climbed back into the engine, and they took off again with sirens wailing, back onto the road. Unbelievable!

Once when we were collecting nectarines in Torino, for Bailey's, there was an ammonia leak in the packing plant. When the fire engine finally arrived, the firemen all ran into the building. Just seconds later they all ran out, coughing and spluttering, rubbing their eyes. It was like an old Charlie Chaplin movie, although not very silent.

For the first three weeks in August, Italy closes down. Everybody, and I mean everybody, goes on holiday. Factories, shops and offices just shut up shop and they all go off to the beach. The roads are packed to the max and service stations on the autostrada are gridlocked. Not a good time to be there.

Italian drivers compete with the Portuguese for the worst in Europe. The truck drivers are chancers and often suicidal. Running through the night, we'd often overtake a wagon with a TV set on the dash, usually showing a footy match, although once it was porn! Not very good for your concentration I would have thought!

Truck drivers reading books and newspapers are two a penny all over Europe. Once just north of Macon, we approached a Spanish Magnum that seemed to be drifting a bit. Cautiously overtaking, I peered across to see a young girl, no more than 12, sat on the driver's lap steering the wagon! The driver was leaning back in his seat, on his phone! Frightening!

Once I really was nearly frightened to death! At the BP Macon truckstop one day, we pulled into the far corner of the lorry park so we could get a few hours' quiet kip. The only other wagon around was an old unaccompanied A-frame drag. I could see it was an old circus wagon, an animal trailer with bars. Curious, I ambled over to have a look.

A metre or so away I froze in horror as a fully grown tiger leapt snarling at the bars! I almost shit myself! My hair stood on end and goosebumps sent shivers up and down my spine at breakneck speed. My fear was primeval. Fight or flight? I could have done neither, I was a quivering, shaking mess.

The big cat calmed down and eventually so did I. I couldn't believe what I was seeing. There was no one about at all. The trailer had just been dropped in the lorry park and left. What if . . .?

Jo joined me and together we gazed at this magnificent creature. It was such an amazing sight, but a truly sad one. Far away from its natural habitat, behind bars to entertain paying punters. In the wild, tigers can have a territory of more than 100 square miles. The cage was about 20' × 10'.

Murfitt's was our first proper double-manning job, well-organised and seemingly respectful of their drivers. We loved the job, running down as far as Taranto, and tipping in Naples and Rome. We told everybody who'd listen that it was a great job. And we were never asked to run bent.

Then, one day after we'd been on with them for 11 months, we parked up late evening in Dijon, France, on our way down to Milan. It had been a long day, starting off with tipping and then reloading in the UK. We parked up with minutes to spare of our 22-hour spread over.

Communications with the office was via a new-fangled, satellite tracking system. Our first ever contact with a satellite, somewhere deep in space. We had a small screen mounted on the dash with a keyboard.

We'd only been parked up a few minutes when a message came through.

'You must be in Milan by 7am.'

To which I replied, 'Sorry out of time. Midday earliest.'

The reply came back in capital letters. 'I REPEAT YOU MUST BE IN MILAN BY 7AM!'

I felt we were being put under pressure. I replied, 'I repeat we are out of time!'

There was no more communication. Although Milan was 'only' eight or nine hours away, we didn't even consider it. We'd gotten out of the habit of being bullied into breaking the law. It came as a bit of a shock to us to be 'ordered' to do so.

We went to bed, ready to do an eight-hour break and get there as soon as we could.

But a couple of hours later there was a knock at the door. A driver had ran up solo from Lyon to take our trailer off us. He had orders to get it down to Milan for 7am. He'd run without a tacho card in all the way. We told him he was being stupid, as we wound down the legs and handed the trailer over.

The following morning the message on the screen just said return to base. We had an uneasy feeling about it as we ran solo all the way home.

A junior office boy, an ex-driver no less, informed us he was 'letting us go.'

'What!' I was incensed! 'Why? What the fuck's going on?'

'You're not working hard enough,' he replied, before stepping back as he saw the anger well up inside of us.

157

'This is our third 22-hour shift this week!' I yelled. 'Why haven't you mentioned this before?'

'Just clear the cab out, will you,' he said, so arrogantly I could have hit him. With retrospect I wish I had. We'd been sacked because we'd refused to run bent!

We tried through the union for wrongful dismissal, but you needed to have served 12 months for that, and we'd only been on for 11. A coincidence maybe?

The scars from that job still lay with me today, and it gives me little pleasure (oh yes it bloody well does!) to tell you that after several high-profile court cases for bent operations, Murfitt's was finally closed down by the Traffic Commissioner in 2003. Everything comes to he who waits!

17

Foggy Freight

FOR the next week or so we sat around the house in a daze, shell-shocked and angry. How can a job that had seemed so good and right for us turn so sour so quickly? We felt we'd been fooled into thinking that we'd found a job where the driver was actually a valued part of the organisation. In the words of that great Who classic, we vowed, 'we won't get fooled again'.

It was good to be free of the tacho for a while. We reintroduced ourselves to the kids and had days out with Kayleigh. We went and saw a couple of bands and just relaxed after so long on the road. But financially and mentally we needed to get back working again.

It was after a month or so of us kicking around the house and making call after call that we struck lucky. A friend of a friend gave us a number.

Freight Transfer ran out of Portsmouth. General haulage, mainly to Spain, Portugal and Italy. Set up in the early 80s by Paul 'Foggy' Fogden with just a single truck, it now boasted more than 25 tractors and around 60 trailers, as well as having more than 20 subbies on the books.

Foggy was a happy cheerful chap. With a BMW and all the bling, he was everything a successful trucking entrepreneur wanted to be. There was a bit of the wide boy about him and when you run with a lot of owner-driver subbies, the rules aren't always followed. If you wanted to run a bit over your time to get home for that day off, well, nobody would pull you up for it.

As the only permanent double-manners, we never felt the need. Indeed, at our interview we made two things pretty clear to Foggy. We didn't run bent, and we needed to have regular and predictable time off. No problem he said. And surprisingly it never was.

So in the spring of 1996 we began pulling for Foggy.

After a couple of runs to Spain, we were given a new Iveco Eurostar. Drivers slagged it off, saying it was Italian and really just a plastic Fiat. But they were usually guys that had never driven one. It was a top motor, especially for double-manning. Loads of space and storage, and you could take your trousers off standing up, the true measure of a good cab. And with 420 horses, she didn't hang about. Great motor to drive as well, comfortable with good vision all round.

With a step framed tilt, we'd run everything and anything to everywhere and beyond. Foggy had a big contract delivering JCBs. Collected from Rocester in Staffordshire and taken to all corners of Europe. Driven on and off, they were the perfect cargo.

Tilts are usually bad news. A framework of wooden slats and iron poles supported the sheet over the trailer. Anything big or bulky meant breaking it all down. Mostly we'd just lift the side for pallets, but occasionally we'd have to strip the whole lot off for slabs of marble out of Portugal, or any of the bigger JCBs. It could take hours of bloody hard work, especially putting it all back together again. And it was always either boiling hot or pissing down with rain.

160

Jo was very hands on, we were a team in every way. It was hard enough with the two of us, it must be a nightmare for single drivers.

We'd often sweat with loads of hand-ball. Shoes out of Italy were always a pain. Just big boxes, stacked floor to roof. Not quite as bad as Pannal and the seat cushions, but a long hard slog. But the guys at the factory always helped and once even gave us both a new pair of boots!

We did a lot of tramping around Europe. One of the longest runs we did started with a ferry to Caen with two JCBs bound for Rho, on the outskirts of Milan. We ran into Italy via the Fréjus Tunnel. We took our first break at Novara services, at the bottom of the mountain, having covered nearly 700 miles in the first 15-hour hit.

Next day we ran an hour into Milan and tipped, before reloading in Reggio for Gibraltar, 1,400 miles away. Dropping down to the coast road at La Spezia, we followed it all the way into France, crossing the border at Ventimiglia. What a road! To our left, the sparkling blue Mediterranean, to our right, steep mountain-sides and forests. And below us, tiny coastal villages and rugged bays. In just 60-odd miles we ran through 102 tunnels and over 108 viaducts – I was counting them for a magazine piece I was writing. Not for those wary of heights or confined spaces.

The descent into Nice is 6 per cent for 10 ks or more and trucks are limited to the nearside lane with a 50kph speed limit. With only a light load and a good exhaust brake, it was no problem at all for us. But just in case things did go wrong, there were five escape lanes on the way down, sand-filled traps that always seemed to have tyre tracks leading in. Scary.

We crossed into Spain at La Jonquera and continued to follow the Med along the Autovia del

Mediterraneo. By-passing Barcelona, we stopped at Tarragona after a non-stop 15-hour hit, having punched out another 700 miles.

The next day we passed the huge Roman fort over-looking the Gulf of Valencia. We now had the Sierra Nevada separating us from the Med. Snow-capped, they seemed to defy the heat of the day.

At Lova we split back down to the coast at Malaga, seeing our first traffic light for 1,200 miles. And yes, it was red! We joined the dreaded N340, supposedly the most dangerous road in Spain – although any road with Spanish drivers on it has to be considered dangerous, and I've seen a lot worse. That said, passing through Marbella, the road – a fast dual carriageway – is like a bloody racetrack. With lots of tiny slip roads on and off, and with cars and mopeds all too willing to pull out in front of 40 ton trucks, it certainly makes you pay attention.

We were too late to make Gibraltar, the border closed for trucks at 4pm, so we stopped at Estepona, 15 or so miles short.

We ran in the next day. We'd done nearly 2,000 miles since leaving Caen, and the Iveco had clocked up her first 100,000 Ks, giving us a good overall 9.1mpg. Not bad for a plastic Italian, eh?

We ran a lot to Gibraltar. Great place to be week-ended. It's like a little England. Union Jacks everywhere, with pubs and fish and chip shops on every corner. And cafés that have English newspapers and serve eggs on toast. Paradise!

The next day we headed out to Granada and collected a load of tiles for Birmingham, nearly 1,500 miles away.

Foggy got us a trip down to Athens. Greece was a first for us. We ran with a couple of JCBs down to Ancona in Italy, then took a 24-hour ferry to Patras. More like a cruise-ship really. Swimming pools, entertainment all day long and with free meals and lodgings.

The ship's docking was delayed a couple of hours by an earthquake in Athens. The following day on the 200K run into Athens, we felt several mini shocks. The earth really moved for us.

We crossed the Corinth Canal and stopped to check it out. It's only four miles long and just 80' wide. When large ships squeeze through there's only a few feet either side to spare. It connects the Aegean Sea to the rest of the Mediterranean. Unfortunately, due to the earthquake, the canal was closed to sea traffic.

In Athens people were out on the streets and in one park we saw lots of pitched tents and first aid stations. There were no fatalities but lots of injuries and structural damage.

At our tip, on the outskirts of the city, all the staff were outside. Their three-storey office block had a huge crack running down the side. The lift was out of bounds, so a few brave souls ran in and out with computers and files.

We got a return load of money. Chewed up Greek drachma in 50 gallon drums, bound for the mint at Llantrisant, in Wales, to be melted down and reused. More than 15 tons of cash.

As we waited to board the ferry back to Italy, we saw some half dozen illegal immigrants, North Africans, attempting to sneak onto a truck. But they were soon spotted and apprehended by the local police. They were forced to sit on the ground, hands on their heads and cross-legged. A big burly officer then proceed to hurl abuse at them, waving his fist in front of their faces and treading on their ankles. Then he stopped, waved his arms and they all got up and ran away as fast as they could.

We'd often see similar scenes in Calais. Families with children, penniless and homeless, trying to make their way to a better life. Robbed and deceived by the people traffickers, cursed and despised by truck drivers. If any manage to smuggle aboard your truck, fines of £2,000 per illegal immigrant are levied.

You cannot help but feel for these people. Fleeing war and poverty that we in the affluent West can only imagine. If we were in their shoes, would we not be risking all to better ourselves and feed our families? Maybe that's too simplistic.

18

Agony in Spain

WE had a bit of an emergency one trip, when we were parked up on our way home, just north of Madrid. I woke up feeling shit. I had a bit of a temperature and some terrible stabbing pains in my side. Getting out of the cab, I fell to the floor, unable to get back to my feet again. The pain was just incredible, I was crying out, yelling even.

Jo called a taxi and I was rushed off to the local hospital. It was kidney stones, and let me tell you, the pain was unbearable. Ask anyone who's been through it and they will agree, it's the worst. Some say worse than childbirth.

Jo had to hold her hand over my mouth as I screamed obscenities and worse. If you'd have placed a gun in my hand I would have gratefully placed it to my head and pulled the trigger. I chewed on my arm and drew blood.

The nursing staff had to hold me down as I kicked and swung out in sheer pain. Just recalling these events brings me out in a sweat. The hospital eventually jacked me up with morphine and my pain was replaced by euphoria.

Jo drove us back to England while I lay on the bunk, with a stupid grin on my face and a bottle of Dr Feelgood in my hand. The pain kept returning in waves, kept at bay only by the morphine.

Arriving back in the UK on a Saturday morning, I

fully intended to visit my doctor on the Monday. No chance. On the Sunday morning hell returned with a vengeance. Daniel rushed me off to A&E, screaming all the way. Again Jo had to gag my mouth as I expressed my desire to die.

At the hospital the waiting room was full of wimps with sprained wrists and bumped heads. I felt I deserved some priority, and told everyone so. I had to be X-rayed to confirm my diagnosis, before they'd give me some more morphine. I must have sounded like a junkie, demanding they give me some drugs, and quick!

Over the next few days I suffered several relapses. I drank gallons of water every hour, trying to flush the stones through and eventually the pain stopped, never, touch wood, to return.

Kidney stones are very rare in vegetarians, but obviously not rare enough. Eating a lot of meat, overdosing on protein, can bring them on, as can too much calcium in the system. Dehydration, not drinking enough liquids, helps them form. Then the stones block the urine from getting to the bladder and your pee backs up into the kidneys. Believe me, that was the most pain and suffering I have ever experienced.

While I was recovering, Foggy sent me on an ADR course. ADR stands for International Carriage of Dangerous Goods. No, I couldn't work that out either. Shouldn't it be ICDG? New contracts meant our general haulage loads often contained nasty chemicals and we drivers needed to know what to do if things went wrong.

It nearly didn't happen, as Foggy insisted that I sign a bit of paper promising to refund the cost of the course should I part company with him in less than 12 months. As no extra pay was involved, and the loads would obviously mean harder work and even more paperwork and rules to follow, I couldn't see any benefit for us at all, let alone end up paying for the course myself. So Foggy had to back down before I agreed to do it.

I did a three-day class in Southsea, run by two brothers. They gave the lessons, then gave you a test, then marked your papers before issuing you with a pass certificate. These were early days of driver education and anybody could set themselves up as a training school.

Alarm bells rang on the first day when the instructor slagged off women truck drivers for pinching men's jobs. If Jo had been there she would have slapped him, but the rest of the class seemed to agree and I was very mindful of who was marking the papers. Then he told us that we were supposed to build a fire in the yard outside and get some practice with different types of fire extinguishers. But as it was raining, we were going to give it a miss. But if anybody asked, we'd done it, OK?

Worse was to come on the last day, just before the exam. The instructor was telling us about an incident in the USA where a fuel tanker had an accident on a hill. Fuel began to leak out and ran into the drains. It flowed through the system passing under a factory nearly a mile away. A tremendous explosion killed 30 people.

'But it wasn't that bad,' he said with a smile on his face. 'Most of them were niggers!'

What? You could have heard a pin drop! A shocked, stunned silence. The whole class just looked around at each other as if to say, 'Did he really say that?'

The silence seemed to embarrass him. Then somebody sniggered, and the instructor laughed out loud, then began handing out the exam papers.

I really didn't know what to do. I know what I

should have done, but I didn't. I was so shocked I just went along with it all. I felt morally guilty for not taking him to task. Indeed, I felt ashamed, and I still do to this day. To say that I needed that pass certificate for the job is not an excuse.

When I told Foggy about it, he told me that the rest of the drivers were going through the same course, and as he'd paid up-front, well, not a lot he could do really, was there? I had no right to criticise him when I'd taken the same option.

Don't let anyone tell you that racism in the haulage industry is on the wane. It's not. I come across it on a regular basis.

One of the first loads of poison we ran for Foggy was paraquat, a toxic chemical used as a weed-killer. You could only carry 15 tons of the stuff at a time because it was so lethal. We collected it from ICI on Teeside. Acres of chimneys belching out yellow smoke, and the bitter smell of the plant was evident for miles before we got there.

We ran the stuff to Lisbon. On the way down, in the early hours of the morning, just before Santarem, we ran alongside a huge reservoir. The road was narrow and had no crash barriers between us and Lisbon's main water supply. If we'd had gone over the edge and disappeared beneath the waves, the whole of the city would have died within days. Makes you think of your responsibilities, doesn't it?

We loved Portugal. Our first runs down there with Bailey's are good memories. With Foggy we did a lot of running down to the south coast, along the Algarve. We'd often carry a few pallets of very English goodies, dog food, baked beans and loads of ketchup. Delivering them to a small farm just outside Faro, where an English guy called Roger, would knock them out to all the local ex-pats, living the life of Riley in the permanent sunshine.

Roger was a real entrepreneur, he had lots things going on. He bred ostriches and was also the local estate agent for Brits coming and going to the region.

He also had a cautionary tale to tell. While living in England, three marriages had left him with five daughters and near bankruptcy. Looking ahead, he flew to the Algarve and purchased a plot of land. He hired a builder, instructed him to build the home of his dreams and then returned to England, working hard to pay for it all.

He got regular reports and updates from the builder and duly sent out more money to get it finished. Then he sold up in England, put all his belongings in a caravan, crammed his new wife and five daughters into the car and set off for his Utopia.

You've seen this coming haven't you? He didn't. Not until he drove into his field and saw, not a villa, but just a pile of breeze blocks and a few holes in the ground. The builder had emigrated to Mexico, no doubt to build his own villa there.

Roger was devastated, as you can imagine. But he got over it. 'I had to,' he said. 'What else could I do? We couldn't go back to England.'

For the next three or four years Roger worked his nuts off, wheeling and dealing, and eventually built himself his villa. Now his businesses were thriving.

There must be a moral in that story somewhere.

While delivering to his farm one day we shit out on the tiny track that led up to his place. I got too close to the edge of the road and our poor Iveco's front wheel sank in the soft mud. At first I didn't think the situation was too bad. But the more I struggled to free her, the more she sank down. Roger brought out a tractor, hooked us up and pulled. No joy, just the threat of him ripping off the front of the truck.

Several locals appeared with shovels and, in the midday heat, dug out around the front wheel. But still she wouldn't budge. She was well and truly grounded, up to the diesel tank on one side.

But everybody was cool about it, really laid back, it was like a day out for them. After slogging for a couple of hours, a break was suggested, and we all retired to the local bar. Two hours later – yes, two hours – we returned to the task in hand. After another hour or so of frustration, a JCB was commandeered. After an unsuccessful attempt to push out he hooked up to the tractor on the front end and together, along with a lot of welly from me, they managed to pull our poor old Iveco free. He pushed from the rear as Roger's tractor tugged from the front. A huge cheer went up from the locals as she finally came free. The only damage to her was a broken side-light. We'd been stuck there for more than six hours. The holes that we left behind took a while to fill in as well.

Over the years we'd had quite a bit of hassle in France and Spain with general strikes. I've always been a union man, the right to withdraw your labour is a fundamental right that people have died for. That right is the only weapon the working man has today.

Us UK drivers often admire our continental brothers for sticking together and acting as one. They'll often close down the country until they get what they want, and most of us support them all the way. Pay and conditions for drivers in the UK are crap, but our unions are not strong enough, or well supported enough, to get results.

But there is also a very dark side to strikes, as we found out in the summer of 1997. We were running to load a curtain sider in Santander, in the Basque region of Spain. We'd heard from other drivers that there was going to be a truck drivers' strike, a big one, in the next few days. Bosses and union reps had been slagging each other off for days and no one was backing down. A strike was imminent.

The following morning our windscreen was replaced, courtesy of Gerposa. Pedro seemed surprised at our lack of gratitude. We had a load booked out of Gerposa, a large forwarding agent. Car parts for Nissan in Tyne and Wear. We parked up overnight alongside their warehouse, ready to load the next day.

But I knew it was going to a bad day when we discovered five flat tyres on the trailer the following morning. They'd been spiked!

Gerposa couldn't have been more helpful. Pedro, their transport manager, apologised profusely for his countrymen's actions, and organised rapid repairs by Gerposa's fitters. Too rapid if you ask me. There were more than 30 punctures in the five super single tyres, most likely the work of a sharpened screw driver. Three fitters took less than an hour to plug and patch every hole. Quite worrying really.

There was an urgency about Gerposa. They wanted us loaded and up the road quick. Pedro assured us that the strike would not begin before the weekend. The flat tyres, he said, were just a few mavericks jumping the gun, trying to intimidate Gerposa's non-union drivers. They loaded us double-quick and, after making a few calls, Pedro told us the road to the border was clear: all 135 miles of it.

What we did not know was that there was a heavy penalty clause on the load for late delivery. Pedro knew the strike was on and the problems we'd face, but hey, that's transport managers for you. We were only the drivers.

As soon as we hit the motorway we knew we were in trouble. Big trouble. Strikers with banners stood on bridges, shaking their fists and spitting. It wasn't long before a car appeared alongside us, full of very angry people. One made the shape of a gun with his hands and pretended to shoot at us. Then they pulled in front and brake-tested us. We slowed and changed lanes. The motorway seemed eerily empty.

The car was soon joined by two others, and after shouting to each other, they all drove off ahead. Not for a moment did we think that was the end of it.

Coming over the brow of a steep hill, we saw half a dozen strikers, stood in a line right across the carriageway. Some had clubs, others bricks and stones.

In retrospect, perhaps we should have called it a day, but I was pretty pissed off by the way we'd been conned by Gerposa, and by the way our fellow Spanish truckers were treating us. First the tyres, now this.

Moving into the middle lane, I gunned it. The guy stood directly in my path must have been a matador. As the others scattered, he stood firm. At the last possible moment (I could see the whites of his eyes) he dived clear, lobbing his brick at the screen. It shattered, spraying slithers of glass into the cab.

This was crazy, someone was going to get hurt. I pulled over onto the hard shoulder. I had little choice. The border was more than two hours away, no way could we make it in one piece. The strikers surrounded the cab, whooping and jeering, like primeval hunters bearing down on their quarry. We tried to reason with them, but all we got was threats and abuse. No way could we get out of the cab.

The cavalry arrived in the shape of the Guardia Civil. Never have I been so glad to see the law. After shaking hands with the strikers, they ordered us to follow them to a holding area back in Santander. The strikers formed a convoy with the Guardia Civil, blasting their horns and waving banners from their windows. It was almost like a carnival procession, but not really.

When we arrived at the holding area, a cattle market, it was full of trucks from all over Europe and Scandinavia. Many showed signs of battle damage and most drivers had their own horrific stories to tell. They'd all been forced to park up.

The threat of violence from the strikers was very real. During the strike, two truckers and a striker died in clashes and several trucks were burnt out.

I tried talking to the union leaders at the site. We're all in this together, I said. We should be united, not trying to kill one another. They didn't want to know, they were very aggressive and wouldn't even discuss it.

The holding area was like a refugee camp. Trailers were opened up and barbecues and chairs appeared from nowhere. There was a small café on site, but most drivers cooked al fresco. It seemed we were in for a long one, rumours abounded of a month or more at least.

On the third day Pedro appeared. Surprise, surprise.

'You lied to us,' I said, not at all friendly like. Jo wanted to slap him, and told him as much.

'No, no, no!' he said, smiling and grinning, trying to defuse the situation.

He'd sorted it all out, he said. No problem. He'd arranged with the strikers to allow us to take the truck, and his precious cargo, back to Gerposa's warehouse.

By 'arranged', he meant that Gerposa had contributed a sum of cash to the union's coffers! Who in turn would escort us back to Gerposa. We were being traded like commodities.

The unions sold their soul for monetary gain from their capitalist enemy, and non-union Gerposa had sold theirs by paying the ransom money.

We ran back to the warehouse in a convoy of horn-tooting strikers with Pedro's car out in front.

Once we'd pulled inside the huge shed, after the strikers had left, Pedro asked us if we'd run the load to the airport, about three miles away. There the load would be trans-shipped and flown to the UK. Of course, we'd have to do it in the dead of night.

He seemed surprised when we said no. Well actually, we told him to fuck off! He'd lied and conned us from the very first moment we'd met.

Gerposa offloaded the trailer, with no help at all from us, then reloaded it on several smaller wagons. They pulled out in the early hours of the next day.

We settled down to spend the rest of the strike parked up inside the warehouse. We'd been stuck there for a week or so when Foggy called for a chat. Now we'd always enjoyed working for him, the work was good, the pay not too bad and we certainly liked the motor. We knew he was struggling a bit, we no longer had the JCB contract and Foggy's BMW had gone, repossessed some said. And the phones had been removed from our cab the last time we were home.

But we were still taken by surprise when he explained the new 'rule' to us. He'd lost a lot of money due to strikes in the past year or so, and he told us he had to make cuts. He could no longer afford to pay us just to sit around in sunny Spain, waiting for the strike to end. So he'd decided to pay us just £25 a day while we were strike-bound. He assured us that as soon as it broke, and we started running again, we'd be back on full pay.

Oh well, that's all right then.

I asked him where he wanted me to leave the keys.

174

'What keys?' he asked.

'The keys to the truck,' I replied. 'We'll be flying home tomorrow!'

'You wouldn't,' he said.

I hung up and waited an hour or so before calling back. When I told him we were booked on a flight that evening, he had no idea if I was telling the truth or not. It didn't matter, he came to his senses and said that rule would only apply to future strikes.

While we were strike-bound I managed to stay in touch with Dave Young, the editor of *Truck & Driver*. I visited various holding sites and filed regular reports about how the strike was going and how all the trapped drivers were coping.

Then I had a call from the *Daily Telegraph*, and I filed a report with them. Another call had me taking a cab to a Spanish TV station and doing a piece to camera for BBC South Today. I got £100 for that interview!

Every other day, Jo and I would catch the bus into Santander for a meal away from the truck and a chance to see the city and stretch our legs. We sat in Santander for 15 days before the strike ended and we managed to run for home.

Our relationship with Foggy changed, it had to. He was stressed out a lot and upped the pressure. We began to get less time off and then he wanted us to run a few solo runs. We declined – that's not what we were about.

One day we came into the yard to find out that the Eurostars were all going. End of contract, said Foggy. Looked like repossession to me.

In their place, four new DAF 95XFs arrived. Of course we were pleased to be given a new motor, and it really was a great truck. But we missed the Iveco and although the DAF was called the Spacecab, we'd had a lot more room in our plastic Italian. She took a while to bed in, but we were soon getting nine–plus to the gallon, and she could really fly up those slopes.

Foggy opened an 'office' in Caen. In reality, it was just a phone in the corner of someone else's transport company. We began just ferrying trailers back and forth across the channel. The job was changing fast.

We started meeting up with a lot of French and Spanish drivers, now apparently subbying for Foggy. It was good to mix with these guys. We saw how they viewed the haulage business and heard their thoughts on driving in the UK. Of course they hated the motorway service stations, and laughed at our so-called truckstops. Paying for parking and showers was just outrageous, they said. None of them liked driving on the 'wrong' side of the road and thought we should change over, now we were in the EU and all.

When we complained about their corrupt police forces, they countered with tales of paying off British police. They gave us several examples of our boys in blue looking the other way over speeding and tacho offences, in return for cash or ciggies. Not too hard to believe, they were very convincing.

The end of the Freight Transfer episode in our lives was eventually forced upon us. We'd travelled up from Spain with a load of shoes for Wolverhampton, puzzled why Foggy wasn't answering our calls. At Caen, Brittany Ferries wouldn't let us board. No ticket, Foggy owed them a few grand and wasn't cooperating in settling the bill. We had the option of travelling as foot passengers, but the truck stayed on the quayside.

As professional drivers we knew we couldn't do that. We felt some weird, guilt–like responsibility. Besides we'd never be able to carry all our gear as foot passengers, we'd have to leave most of it behind in the motor.

We managed to contact another of Foggy's drivers and the news wasn't good. Tractors and trailers were being repossessed in the yard as we spoke. Foggy was lying low, incommunicado.

We phoned the shoe customer and explained the situation. Maybe it was something I said, but he pretty soon realised that if Foggy had indeed gone bust, he could well lose his load of shoes to the liquidators. I must admit, I led him along that path, but we were desperate to get home. After promising that we'd deliver the shoes before we went back to the yard, the guy paid for the truck's passage.

We did as we promised and delivered the load to Wolverhampton, getting back to our yard by late evening. A couple of guys sat in a car in the deserted yard. They'd been waiting for us all day. They were from the trailer hire company and were glad we'd made the delivery, now they could have their trailer back without any hassle.

They were pretty cool and, like us, were just doing their jobs. We even ran the trailer over to Southsea for them, back to their yard. Several of our trailers were there, in various stages of having the company name stripped off the sides.

We returned solo to our yard. The office was all locked up and the yard was empty except for the old shunter. Before we'd finished clearing out all of our gear, two men in car arrived to take over the truck. The trailer guys had let them know we were back.

A sad end really to a good job. Foggy started up again a month or so later, much to the anger of the subbies that he still owed thousands to.

We had a week or so off, making a few calls to firms we knew did double-manning, but had no luck. Then a friend put us in touch with Pro-Trans, a three-truck outfit running out of the truckstop at Farlington, just outside Portsmouth.

A Portakabin in the corner of the lorry park was manned by a lad called Dave. In his late 20s, we were warned he'd already been bust a couple of times and could be a bit of a tyrant. But we needed the job and were pretty confident we could cope. We'd had lots of experience in dealing with arseholes. We decided to give it a try and we got the job.

We told Dave our mantra, we didn't run bent and we wanted regular time off, every two or three weeks or so.

'No problem,' he said, so quickly that we just knew it would be.

The wagon was an Iveco Eurostar, great! With an Eaton Twin Split. Shit! We both hated them with a vengeance.

And yes, he was an arsey git, but we didn't let him bully us. Well not too much. That first trip, Spain, Portugal and France, kept us away from home for seven weeks! Every time we got back to the ferry at Calais or Caen, we'd swap trailers in the docks and run back down again.

We demanded our time off and eventually headed back home for a well-earned three-day break.

Back in the yard, as we dropped the trailer, Dave ambled over.

'That trailer,' he said, pointing to a tilt in the corner of the yard, 'has to be in Macclesfield by ten in the morning.'

'We're on time off,' I said, firmly.

'Ten o'clock,' he said. 'Or you can just clear out your motor and piss off!'

'Oh, OK then,' I said.

He walked away with a swagger. He was used to getting his own way.

I think he may well have got the wrong impression when I said OK. I meant we'd clear out the truck. When I slapped the keys and fuel cards on his desk he was gobsmacked.

He tried to backtrack. He'd deliver it himself, he said. 'See you next week'.

He needed us more than we needed him. Thanks, but no thanks.

I don't believe in much, but I do believe in karma. The driver who took on our job laid the wagon over on a roundabout, just outside Southampton, a few days later. He was unhurt, but I bet Dave's pocket wasn't!

19

Scandinavian Doubles

AFTER seven weeks away, it was good to have a bit of free time. Waking up in the morning, after more than the usual eight-hour break, it takes a second or two to realise you're not in the cab. Walks with the granddaughter and visits to a few country parks helps you get out of the cycle. But we did need to work.

We phoned around a few firms, just trying our luck. When we tried Europa, a large continental haulage firm from Erith, they couldn't help us but suggested we try one of their sub-contractors, Eurospan.

I spoke to their transport manager, Stuart Kirk, early one morning, and he suggested we go see him for an interview. After telling him our history, and giving him a few references so he could check us out, he suggested we bring along our gear, just in case. Would we be OK to ship out to Sweden that night? Music to our ears!

As we quickly got our gear into the car and set off, we had one nagging worry. We'd given Bailey's, Murfitt's and Foggy as a reference. Foggy was no problem, if not for the fact he was now only using single-manners, we'd have rejoined him. And Bailey's weren't too much of a concern, but Murfitt's, well they could be. I was glad I hadn't hit the little tosser!

Eurospan ran half a dozen tractor units, subbying for Europa, who had depots all over Europe and

Scandinavia. They had one other double-man crew, two guys, and did regular timed runs down to Italy and Sweden. Their office and small yard was just around the corner from Europa in Erith, close to the Thames, a two-hour hit from home.

When we entered their small office, I could see straight away that there was a problem. Stuart had spoken to Murfitt's and wasn't keen on what he'd heard.

We sat down and over a cup of coffee we explained, trying not to show too much venom, how the job had ended so badly.

Stuart surprised me. My mantra of 'all transport managers are bastards,' was blown right away. He listened to our story, then confessed he'd heard tales from other drivers on the way Murfitt's ran. In fact they had undercut Europa on many occasions and taken work away from them. It's easy to do the job cheaper when you ignore the rules.

The tables were suddenly turned. Instead of us promising to work hard and do our best, it was Stuart promising us he'd never ask us to run bent and telling us he'd guarantee us time off on a regular basis.

And I will say this now, that in the three years or more we worked with Stuart, he was true to his word and without a doubt the best transport manager I'd ever worked for. Caring and compassionate, he saw drivers as real people with their own lives to live, not just as expendable steering wheel attendants.

So in the late 90s we began running regular trips to Sweden, pulling Europa trailers. Non-stop, London to Helsingborg in 22 hours. Driving for about 19, the other three boarding and sailing three different ferries. After a few months we were given a brand new DAF 95 Spacecab. It was just perfect for the job, and a pleasure to drive with its ZF16 gear box.

In Helsingborg, we'd back onto the bay at Europa's Swedish depot, and take our eight-hour break while the curtain-side trailer was tipped and reloaded. Then we'd run more than 500 Ks up to the Saab plant in Stockholm, tip, reload and drive back down to Helsingborg.

Sweden is such a beautiful country. Huge mountain ranges and vast lakes, and with roads just perfect for eating up the miles. Eagles and ravens fly effortlessly in the sky.

We often saw moose – thankfully behind the high fences that keep them off the road. A moose is about the size of a carthorse and they are responsible for hundreds of deaths a year on the roads of Scandinavia. When a car hits one it takes away the legs from under it, and the whole body of the animal just smashes in through the windscreen. Gruesome pictures appear regularly in the press and on TV. Bullbars are an essential defence on trucks over here, not just the cosmetic extras that you see in the UK.

On returning to Helsingborg from Stockholm, we'd take another eight-hour break at the depot, while the trailer was again tipped and reloaded, then one-hit it back to London. After a couple of days off, we'd do it all over again.

The trip had us passing through seven countries in one day. Out of England at Dover, up through France, Belgium and Holland. Then north up through Germany, a ferry into Denmark then another over to Helsingborg in Sweden. Not far short of 1,000 miles in one hit. We'd only ever stop for fuel, ferries and changeovers. By plane it's less than two hours.

It was a great job and we loved it. We got into a driving pattern that suited us both. I would drive down to Dover and board the ferry. Then in Calais, I'd pull her out to the car park where Jo would take over and do a four-hour hit into Holland. I'd sleep solidly in the bunk, confident that I was in safe hands. After a quick coffee and a pee, usually while fuelling up, we'd change over and I'd drive for another four hours as Jo slept. We'd repeat this pattern all the way to Helsingborg and Europa's warehouse, just a few hundred yards from the docks.

We were earning good money too. A proper wage each meant we were taking home our best pay yet. Over £700 between us.

For a bit of variation, we'd do the occasional Italy, usually via the Eurotunnel. Not much had changed, we still preferred the ferry.

A bit of UK running reminded us just how shit the conditions were in the UK. The hardest work we did for Eurospan was in the UK.

On one of our Italian trips we met up with some Persians when we were weekended at Gobba services, to the west of Torino. We found the whole lorry park full of old Yankee trucks, there were Macs, Kenworths and even old Fords. They all had Iranian plates, it was like a transit camp.

The drivers had erected canopies to shade from the sun, and some of their trailers had been opened up with cookers, chairs and small TVs set up. One of the drivers had studied in London and spoke very good English. He explained that they had travelled the 3,000 miles from Tehran to deliver cloth and carpets. Now they just sat there, waiting for a load back home. Some had already been waiting more than three weeks. They wouldn't consider running back empty.

They all owned their wagons. None had covered less than a million miles, even before the Iranians had bought them, through a South American dealer.

We spent the evening sat in a small circle in the

lorry park, drinking their strong neat coffee and discussing the world. Our interpreter was very articulate and as eager to find out our views on the world as we were theirs.

Back home we had 'joy riders' to watch out for, in France it was 'ghost riders'. We met up with some in the early hours on the Reims ring road. We'd just done a change over at the péage and set off with me behind the wheel. Traffic was light, as always at that time of night. Ahead I saw a set of headlights coming towards us. I assumed it was on the other carriageway, as you would. Then when I realised it was actually in the outside lane and coming towards us, I began to flash furiously. The second wrong assumption I made was that the driver had made some terrible mistake and was unaware of his predicament. It was when he swerved at us and tried to put us off the road, that I realised we were part of his night's entertainment.

It was all over in seconds. He missed us and we carried on, thinking 'Did that really just happen?' We later heard that this was the latest craze among thrill-seeking degenerates around France. Steal a car and run the wrong way down the autoroutes. What fun!

One Friday night, heading for home, a blizzard hit the mountain as I climbed up towards the tunnel at Mont Blanc. Visibility was down to fuck all and we were struggling to keep moving with a very light load on board. I woke up Jo, two sets of eyes are better than one. The wheels began to spin on the snow-packed road and we eventually ground to a halt on the side of the road. At low revs I managed a yard or two, then she stopped again. The diff lock made little difference. The spinning wheels seemed to clear the snow and we'd crawl a few more yards before stopping again.

My immediate concern was for any following traffic ploughing into the back of us. I'd been running with the fog lights and hazards on, but they were like candles in the swirling snow. I thought of Teapot, what would he do? Probably put the kettle on. But no, this was serious stuff, we were in imminent danger.

Suddenly, all around us, the snow turned a bright pulsating yellow. A snow plough slowly appeared through the blinding blizzard, coming down the mountain. The driver saw our predicament and actually did a three-point turn in the road behind us. Then, drawing in front of us, he began spraying the road with grit. He reversed up to us so close I thought he was going to hit us. I quickly got the message and began rocking the DAF gently until the tyres began to bite. The snow plough pulled slowly away, and as we got to the grit we began to steadily move forward. The snow plough stayed just a few yards in front of us for the next ten miles or until we made it to a garage, just at the bottom of the last climb for the top. When our saviours saw us turn onto the forecourt, they beeped their air horns and went on their way. Our heroes. They had bailed us out of a very sticky situation and we were eternally grateful.

The garage forecourt had been cleared of snow by a little mechanical snow plough. But if we thought our problems were over, then we were definitely wrong. I pulled onto a pump, solely because there was nowhere else to stop. A few trucks sat deep in the snow around the edge of a small car park. I was just so relieved to have got out of jail free, I just turned off the engine and sank back into the seat. Yes, we'd got away with that one!

A knock on my door focused me. Yes? A little Italian in overalls was speaking quickly and very loud.

An excitable lot these Italians. He kept pointing back out to the road.

'I think he wants us to leave,' said Jo.

'Andare avanti!' he kept saying, waving his arms about. 'Avanti! Avanti!'

I got the message. He wanted us off his forecourt. It wasn't as if we were keeping his customers away. The weather was horrendous, a full-blown blizzard. Nothing was moving out on the road and we certainly weren't going to buck that trend.

Then he started shouting, 'Polizia! Polizia!'

I wound up the window and Jo put the kettle on. We were going nowhere.

About an hour later, as we watched the storm from the heat and safety of the cab, another knock on the door woke me from a light doze.

The polizia had arrived in the form of a very wet and cold looking young officer.

'Ciao,' he said. 'Problem?'

'Big problem,' I replied. 'Neve! The snow!'

I got out and shook hands with him, then tried to explain that the severity of the storm, plus our very light payload, meant we were a hazard on the road. Besides, apart from the snow plough running up and down past the garage, we'd not seen another truck on the road since we'd been there.

He explained he'd come down from his station at the tunnel in his jeep, and that the tunnel was closed. It was even worse on the French side. He was very understanding and he told the attendant that if he wanted us off his pumps, then he'd have to clear a space in the car park for us with his baby snow plough.

Which is exactly what happened. Common sense prevailed. We got our own slot in the car park and there we sat for three days! The storm actually got worse before it got better.

When I phoned Stuart, just after we'd called it a day, he reacted in a way that would shame most of the transport managers I've worked for.

He simply said, 'Mick, you stay there until you think it's safe to run.'

Next-day deliveries or not, our safety always came first with Stuart.

When we eventually pulled out and got to the top, they wouldn't let us through the tunnel because we had a metre or more of snow on the top of our trailer. A flat-back truck with a cherry-picker on it arrived and a guy went up with a broom and cleared the lot off for us. All cheerfully done and in good humour.

Coming down the other side into France was bloody awful. Memories of Pamplona and Teapot played in my mind as we exhausted it round hairpin after hairpin on hard packed snow.

The road's edge was just made visible by snow poles jutting out of the snow. At the bottom a queue of trucks waited by the customs post. The French were still not letting anyone go up. Perhaps they were waiting to see if we made it down. On the CB, several Brits said they'd been waiting for four days. We also heard that the rest of France was snow-free. We banged on for home, better late than never.

Things went well at Eurospan, although we should have known it couldn't last. For reasons unknown, we lost the contract with Europa. The office and yard came with the contract, so we all relocated to a warehouse and yard across the other side of the Thames, in Barkingside Industrial Park.

The job changed dramatically. No more over the water, instead a very hard slog in the UK. Stuart Kirk jumped ship and went to work for Europa. We were sad to see him go, even more so as his principles went with him. We came under a lot of pressure, the work rate went up and days off seemed to be spent in service stations and lay-bys.

We carried anything and everything to all corners of the land. Car parts from Merthyr Tydfil to Milton Keynes, sugar from Thameside to Liverpool. You never knew what or where your next load would be. Proper tramping, it became less enjoyable and more like just a job.

As we were now the only double-manners on the firm, we were expected to do twice as much as the solo drivers. This became very difficult, as we seemed to be the only ones obeying the rules.

How much more can you do in a 22-hour day? We'd often arrive at a tip at the end of our spread-over at maybe four or five in the morning. Then be up to help tip as soon as they opened. This would then end up being our daily rest break, so by midday we'd be running again. It was a good job we were both able to get some shut-eye on the road.

One particular job had us boiling mad. We'd picked up a trailer on behalf of EJ Rogers. We ran it out of Swansea and tipped and reloaded overnight in Glasgow. It then had to be in Peterborough the next morning. Only a double-man team could do it, it's why Eurospan got the job.

On arrival at the plant in Peterborough after a very long night, we queued behind a Bulgarian Willi Betz at the security box.

When it was our turn to check in, the obese tat-tooed guard said to me, 'Tell your missus she'll have to wait outside. Health and safety, no passengers.'

I explained the situation – she was a driver, not a passenger.

'They all say that,' he sneered. 'You can't bring her in.'

I tried to explain. The load was urgent, I said. It needed a double-manned truck to get it here on time.

He wasn't listening. 'You ain't coming in with her in the cab.'

In the yard I saw Willi Betz swinging around up to the loading bay. And it had two drivers in the cab!

'What about them?' I asked, getting angrier by the second.

'They're different,' he said, walking away with an arrogant swagger. 'If she don't get out, you ain't coming in.'

Jo was not very pleased with his attitude and wanted to tell him so. But that would certainly have inflamed the situation, believe me!

So I pulled away and drove down to a butty van a mile or so away. As Jo cooled down by getting the tea and sarnies in, I phoned up our transport office.

They passed on the message to the customer that we were having our break, and if the problem had not been resolved by the time I'd wolfed down my egg sandwich, we'd be taking the trailer back to where it had come from.

Just in time, as I was wiping the remnants of my breakfast from my chin, the call came. The buyer who'd ordered the stuff frantically apologised. He needed the gear now, people had been stood down waiting for us to arrive. I told him we'd meet him at the gate in five minutes.

The barrier raised as we approached. The buyer stood there, waving us through. Jo lent across me and waved at the obese twat. But she only used one finger!

Eurospan developed a groupage partnership with a Scottish carrier and we ended up doing a lot of trailer changes just south of the border. Running our full hours up and down the country.

Some of the loads were Scottish booze for export, tipping in East London. We both thought it was all a bit weird. We'd turn up at the given time to find the supposedly bonded warehouse unmanned. Then two or three cars full of guys would turn up and tip us. Not the friendliest guys we'd ever met. Then they'd close up and leave as soon as they'd finished. This happened twice in a week.

What? Not more crooks? I hear you say. Yes I'm afraid so.

One day, about a month later, we got back to the yard and found to our horror that the owner had been arrested by Her Majesties Customs and Excise on charges of tax evasion and money laundering. The tax-free booze, meant for export, was being sold on in the UK. Millions of pounds, we were told, had been lost to the exchequer.

Eurospan had all its assets seized and was placed under administration. The lady liquidator (ironically from our home town of Worthing) assured all of us drivers that our jobs were safe, and that they were in the process of finding a buyer to take on the whole running of the place.

We believed her because we had to. We were all shocked at the revelations, but things suddenly dropped into place and it became clear just what had been going on since the loss of the Europa contract.

Trailers were repossessed and a few tractor units were sold off. All the time we were being given assurances that we, Jo and I, were safe.

Then on returning to the yard one afternoon, three or four weeks after the shit hit the fan, we were called into the office.

'Clear out the cab,' said the liquidator lady. 'We've sold off your truck!'

We went ballistic! I had to hold Jo back. She seriously wanted to wipe that sanctimonious look off her face. Although I agreed in principle, I managed to keep some control.

'You promised us, you fucking bitch!' I screamed, still holding onto a more vocally stressed Jo. 'You're just a bunch of fucking asset strippers, you had no intention of keeping us in a job!'

I won't dwell on the rest of our conversation. I'm not proud of what I called her, but it was all true. We felt used and abused. We were eventually encouraged to leave the office by a couple of burly security men. They stood over us as we emptied our gear out the DAF, then escorted us off the premises. Again, thanks for the memories!

It was late evening by the time we got home. The kids stood clear, we were still fuming. Yes, we should have seen it coming, but I suppose we didn't want to. We sat with a cup of tea trying to find some positives.

They say that every cloud has a silver lining, well ours certainly did. We'd only been home for an hour or so when the phone rang, it was Stuart Kirk. He'd heard the news and offered us his commiserations.

Then he hit us with, 'Fancy an Italy? Ship out tomorrow afternoon?'

If he'd been in the room I would have kissed him! We snatched his hand off and began pulling for Europa the next day.

20

Simply the Best

WITHOUT a doubt, Europa was the best dou-ble-manning job we had. The truck, a Volvo FH12, was a great truck to drive. Although the cab was smaller than we were used to, she had real power and handled like a dream. We worked hard but were generously rewarded, our best wage yet. We ran set runs with set days off.

A three-day run to Helsingborg and back was fol-lowed by a 48-hour break. Then a four-day run to Helsingborg and Stockholm with another two days off at the end. Then we did a big eight-day Italian run followed by three days off.

It was a three-week circuit run by three double-manned teams, all well-organised and professionally run. Stuart Kirk was running us and my earlier impres-sions of the man proved correct: he was everything a transport manager should be.

The Italian job was by far the hardest work we'd done for a long time. We'd ship out on a Saturday lunchtime and one-hit it down to Chamonix, on the French side of the Mont Blanc Tunnel. Nearly 600 miles with only a pee stop.

We'd have a 24-hour break, waiting for the Sunday driving ban to finish, then in the early hours we'd run through the 11-kilometre tunnel and down to Europa's depot in Milan. Stopping just long enough to swap trailers, we'd then shoot back over the mountain, through the tunnel again and punch three or four hours back up into France.

There we'd meet up with a single-manned driver who'd left the UK at the same time as us. We'd swap trailers, he'd take ours back to Erith and we'd turn around and take his down to Milan. Then we'd take a break before doing it again, only then we'd run back up as far as Macon before the trailer change. Sometimes we'd do a few collections around Milan and Torino.

We spent a lot of our breaks in the road outside the Milan depot. We never slept well. Sometimes these breaks would begin at three or four in the morning, other times it could be four or five in the afternoon. The body clock never adjusted, we both slept better when we were on the road. It was a very dodgy area around the depot and the mosquitoes insisted you kept the windows closed even in the stiflingly humid heat. The mozzies loved me – I would awake with bites all over me, while Jo never got one.

On the Friday night we would hang around the warehouse waiting for the 'Friday night groupage'. Trucks from all over Milan and beyond would ferry in goods for delivery in England on the following Monday. We'd spend hours waiting for maybe just a pallet or two that was always 'ten minutes away'.

We wanted to get going because this was the biggest direct hit of the week. More than 700 miles to the ferry at Calais. And we'd usually been up for most of the day, unable to sleep with the heat and the trailer being noisily loaded.

We'd rarely pull out before 10pm. I'd then take her up and over the Blanc while Jo slept. Then we'd swap over in France and I'd manage a good solid four-hour kip as she ran on.

You had to have a lot of confidence in your driving

partner to sleep like I did. Jo was a brilliant driver, but more importantly she knew her limits. Our absolute bottom line when driving was this: if you feel tired and your eyes are going, wake up your partner! No ifs, no buts – just do it!

Sometimes just a chat and a bit of company can make all the difference and you get through it. Other times, if you need a break, we can change over for an hour or so. There's no shame in pulling over. You don't get a medal for pushing on, you just get an early grave. Without the knowledge that Jo would do this, I would never be able to sleep so soundly as she pushed on through the night in all weathers.

We had a couple of lucky escapes while running the Blanc.

One day we ran through southbound with a subbie. He stopped for a pee at the top as we punched on. When he came out and ran down the mountain he found a huge avalanche had gone across the road and completely blocked it. We never felt a thing, but we can only have missed it (or it missed us) by ten minutes or so.

Then on 24 March 1999, we ran south through the tunnel at about 8am on one of our runs. It was the third time that week we'd run through. Just a few hours later at 10.45am, a fire in the tunnel killed 39 people. It burnt out of control for several days.

The first we knew of it was a frantic call from Stuart when we arrived in Milan. He was really concerned and very relieved to be speaking to us on the phone.

Disasters like that make you stop and think. And not only what a horrible way to die. There but for fortune ...

For a long time we had to run in and out of Italy via Fréjus Tunnel. A longer route, but a much easier climb on both sides of the border.

I loved the beauty of Italy, the countryside and mountains, the roadside Roman ruins, and most Italians we met were always ready and willing to direct us or help us out. But I have to admit, I never really felt safe there. You heard so many tales and stories of cab thieves and muggings. As elsewhere in Europe, baddies know that truckers always carry cash for their expenses, as well as CBs, cameras and duty-free. But they didn't just stop there. We heard several first-hand accounts of thieves plundering the whole cab. They would use the driver's own sleeping bag to carry away all the swag, which would include all your clothes, dirty or clean, all the food, bedding and towels. You come back to the cab and all you've got left is what you've got on you.

We took what precautions we could. I always carried passports, cash and licences in a bum bag that was never left unattended. Running late at night, we'd take our pee breaks one at a time, then have a carry-out for a meal. Service stations and lorry parks could be scary places late at night. We weren't paranoid, just cautious.

On one of our breaks, sat in some services south of Torino, we watched a scam unfold in front of our very eyes. On my way back to the truck with a takeaway, I was approached by a man with a carrier bag. He told me in broken English that he was a truck driver who was low on diesel. So he'd nicked a super-dooper camcorder off the back of his truck and was selling it cheap to get some fuel. So he could get home to his wife and little babies!

He took it out of the bag. It was just as he said. Brand new and very impressive. £50, he said, worth £250 at least. If something seems too good to be true, then it always is. I declined the offer and returned to the cab.

Jo and I then watched as he approached foreign tourists, usually in their cars as they were leaving. It soon became obvious he was working in tandem with a fellow conman. Every time one offered the camera through the window of a car, the other would stand in the car's blind spot and get as close as he could. And guess what? He had an identical carrier bag.

As an unsuspecting punter counted out his cash, the bag with the camera in it was placed on the ground alongside the car. Villain No. 2 then crept on his hands and knees alongside the car and switched the bags. The 'camera' was then passed into the car, cash was passed out and the two highway robbers practically ran off.

The victim began driving out of the car park, then his brake lights suddenly came on, and he reversed quickly back. Leaping from his car, his rage was apparent. In one hand he held an empty camera box, in the other a brick! We saw this scam and a few variations of it, several times. People make a good living out of it.

Beggars are seen at most service stations, Romany mothers with their children, selling lucky heather. Ignore them and they yell abuse or spit. Give them something and you are immediately surrounded by others demanding you give to them as well.

On a trip down to Naples we had beggars banging on the door of the truck at traffic lights. Prostitutes lined the main drag in, and derelict and stripped cars lined the street outside our tip. The security guard at the factory had a shotgun and assured us he wouldn't hesitate to use it if someone came over the wall at night.

We stayed overnight within the citadel, dining with the guard and his wife, who came from Glasgow. She advised us not to go out. We listened to them that knew.

The Italian male has a worldwide reputation as a Casanova and great lover. The truth is they can be a bit of a pest. Jo had long blonde hair and attracted attention both on and off the road. In bars and restaurants, men would just stare and mouth words like 'bella' or 'amour'. She'd stare right back at them and even make eye contact, but they wouldn't flinch.

Out on the road, if I was on the bunk, an overtaking trucker would often stay alongside for mile after mile, blowing kisses and even singing love songs. The moment I sat up they'd drop a gear and shoot off. Most of the time we found it rather amusing and laughed it off. What else can you do?

But one night it was deadly serious. Jo had pulled out of a service station, just north of Milan, with me on the bunk. After a few Ks, a car appeared alongside and the driver was yelling at her to pull over. He kept pointing to the rear of the truck, indicating she had some sort of problem. Then he shot off and was waiting in the next lay-by. Of course Jo just kept rolling. I climbed off the bunk as the guy caught us up again, and again he tried to get Jo to pull over. Then he saw me in the passenger seat and shot off at high speed.

You never know what he had in mind, but he obviously thought Jo was on her own, and silly enough to stop.

The biggest threat we had to our security came when we loaded hanging garments out of Bergamo, one hot sunny day. The plan was to load and run towards Macon for a change over. It was a real pig to load. Hanging garments meant hand-balling arm loads of clothes onto rails, three or four high, down the length of the motor. It was so bloody hot in the back of the trailer that we kept bailing out for fresh air.

I took a call from Stuart, in the factory office, and he asked if we had room for a couple of pallets on the back. No problem, we had plenty of space. I took down all the collection details, a warehouse in Lodi, maybe an hour or so away.

The office Romeo, once he'd stopped flirting with Jo, was very helpful in giving us advice on how to get there. He even phoned up the customer and got precise directions right up to the front door. Cheers mate, every little helps.

On arrival at Lodi, the office girl told us she had a message for us, from our 'colleague'.

'He said he will meet with you in Courmayeur,' she said.

The hair rose on the back of my neck. We had no colleague. No one knew we were here except Stuart and, of course, Romeo back in Bergamo. And how would a 'colleague' know the phone number of the customer in Lodi? Only Romeo knew that.

When I questioned the girl further, she told me that the guy had said 'Tell the English driver I will meet him at the services.'

A colleague that doesn't even know my name?

I phoned Stuart, he agreed it all sounded bloody dodgy. In fact we had little doubt that if we pulled into Courmayeur, the last services before the tunnel, there'd be someone waiting to take the truck off us. Fashion garments are one of the most hijacked loads in Europe. We discussed going back through Fréjus Tunnel instead, or maybe run back to our Milan depot and run up early hours of the following day. Stuart was very concerned, but told us the decision was ours.

We decided to go for it. Run straight through to

the Blanc, stopping nowhere on the way. If they're expecting us to stop – and why wouldn't they? – it's unlikely that they'll give chase as we sail on by. And even if they do, it's only 10 Ks and the roads are pretty busy, there can't be too many places they could pull us over. Once we got to the top, the police and customs presence should deter any hijack attempt.

And that's what we did. We were pretty sure we weren't being tailed, and we just flew past the services at 90 Ks. A bit nerve-racking in the end but we got through and phoned a relieved Stuart from the French side.

Hijackings are such a common occurrence throughout Europe, even here in the UK transport the media regularly reports such incidents. Our experiences certainly brought home to us that you're just as vulnerable as the next man. Stay vigilant and be aware. People are not always as they seem. First impressions can often be misleading.

We learnt another valuable lesson about people in Sweden. Although even now, I'm not sure what it was. We'd been on for Europa for over a year and were enjoying our work. It was pretty demanding but we liked to be stretched, and the pay was our best ever.

The Helsingborg depot was manned by half a dozen warehouse staff and a small office. A crew room had showers, cooking facilities and a TV. Although we usually only spent eight hours there, and then most of it in our bunks, we enjoyed our visits.

We got on well with them right from the start. Handshakes all around on arrival, then the same when we pulled out. They were our friends, we even risked bringing them a case or two of booze occasionally – something that could get you into a lot of shit with customs. Then it all changed in one unexpected and surprising incident.

We'd arrived after our usual big hit to be told we'd be running a different trailer back to London, so it could be serviced. No problem, we'd done this before, and it was a bonus for us because it sat already loaded in the corner of the yard, meaning we'd be able to take our break without the forklift running up and down inside our trailer as we slept. It was also a bonus for the crew, as it was job and finish for them. As soon as they'd finished with us, they were all off home.

I dropped our trailer on the bay and went to hook up to our new one. But on doing a walk around, I found a bald tyre. Not just a bit iffy, completely bald. Whoever had been pulling this trailer must have been blind or just plain careless. It must have been like that for weeks if not months.

On informing their office we were told by the boss, 'Yes we know, that's why it's going back. They will fit a new one in Erith.'

'I don't think so,' I replied. 'Sorry, but there's no way we are running that trailer. That tyre is not only illegal it's bloody dangerous!'

'You will be all right. The tyre is not a problem, just do it!'

Now this was becoming pretty awkward. They were suddenly being very officious and their body language changed. In my mind I could have been back in Lucking's office, or even Bailey's. We didn't have to put up with that sort of bullying any more.

'Sorry,' I said again, firmly. 'We will not be pulling that trailer. I'll put it back on the bay for you if you like, so you can transfer it over, but that's it.'

His voiced raised a pitch or two and he began

babbling in Swedish and waving his arms about. The warehouse foreman was called in, and he too started to verbally abuse me.

'Anybody else would take it,' he yelled. 'Call yourself a fucking truck driver?'

The boss called Stuart on his home number. He told him the situation, then passed the phone over to me.

'Hallo Mick,' he said. 'Is the tyre legal?'

'No,' I said.

'Then don't pull it. They'll have to trans-ship. Put them back on.'

You couldn't ask more from your transport manager.

I left the office and put the errant trailer on the bay. The lads in the warehouse weren't happy and let us know it. They jeered and made gestures that were, to be honest, obscene.

We saw these guys as our colleagues, our friends, but they turned out to be very fickle and shallow. They made as much noise as they could as we tried to sleep. They left a forklift truck running alongside the cab and really slammed all the pallets down, singing and yelling the whole night through. When we finally pulled out it was to jeers and whistles.

As you can imagine, this sudden change of behaviour stunned us a bit. You think you know people, then suddenly a very dark selfish side emerges. In your head you try and analyse the incident. Were we really so naïve as to think they would consider our welfare over the need to get off home early?

Our next run was the Italian trip so we didn't get back to Helsingborg for a couple of weeks. But when we did we realised the job had changed forever. It was like we were now persona non grata. No warm welcome, lots of mutterings and very noisy loading.

But there you go, we'd done the right thing, and I'd certainly do it again.

We enjoyed our time at Europa. Definitely the best double-manning we'd done. But the only permanent thing is change, and the job certainly did. Swedish owner-drivers undercut Europa on a couple of contracts. We lost the four-dayer to a single-manned truck who ran it bent. Illegal and highly dangerous, but hey, somebody saved a few quid, so that's all right then.

We started doing just one Sweden and the Italy. Then that Sweden went the same way and we started getting a lot of time off. We were now the only double-manners left. I occasionally did single-manned runs here and there, but it wasn't what Jo and I wanted to do.

The job had changed and we were ready to move on. We said our goodbyes to Stuart, and told him how well we'd enjoyed working with him. He was a rare good guy in the world of transport management.

21

Back on the Agency

WE took a few weeks off. Had weekends at home for the first time in years, saw friends and watched a few bands. You know, things that normal people in normal jobs do all the time. Waking up in the same place every day was quite nice for a while, but the view out the window never changed. Not heading for or running back from anywhere made us listless and tired.

By now we had a VW camper van, so we took Kayleigh off to Sweden for a couple of weeks. We showed her some of our favourite places and discovered a few new ones. We got a real good close-up of a moose as it ambled across the road in front of us on a forestry road. Driving back down through Denmark we stopped at a bird reserve and watched avocets. Very rare in the UK, this was a first for me.

Back home, we started looking for more work. There was very little double-manned work about. We went for a few interviews but most of them were with outfits that really weren't geared up for it. Most offered the occasional double-manner, but were more interested in two single drivers. That's not what we wanted at all. At most interviews it was us interviewing them, and they always seemed to come up short. Any sign of arrogance or bullying had us heading for the door.

Jo decided to take some time out. Her sister Margaret, in Barnsley, had been diagnosed with cancer and was losing the fight. Jo had already lost a sister, Anne, to this dreadful disease.

I signed on for ADR, an agency in Southampton some 50 miles away. They were a decent firm, with lots of work and a good attitude to their drivers. The money was pretty good too. In those days, agency drivers were paid more than employed drivers. A quid an hour more at least. But with no sick pay or holiday pay and no guarantee of work, employers were still getting a good deal.

I began pulling containers out of Southampton Docks for Securicor. I'd drive down on the Sunday night and sleep in the truck. Then work through until the Friday night before returning home.

Securicor had moved into the container business when it bought out Russell Davies a few years earlier. They ran turquoise Renault Premiums, Scanias and a few Mercedes. I usually had a different motor every week.

The work was constant, week in, week out. Mostly tramping around, only returning to base once or twice a week. I'd do a lot of 3am or 4am starts and work through until 6pm or 7pm. It could take three or four hours to tip a box, so there was always time to catch up on your sleep. Drivers were not expected to assist with loading or offloading. We just delivered the container, full stop.

Attitudes towards agency drivers, from the bosses and drivers alike, were often hostile. To the drivers, we were pinching their work and getting paid more money than them. To the transport managers, we were the low-lifes who trashed their trucks. Any damage to the wagons was always blamed on those 'bloody agency drivers'.

The role of 'driver assessor' hadn't been invented back then. You turned up at the given time and you were just given some keys and told to hurry up. I never left the yard without a good walk-around and any damage reported to the office.

They had no garage or mechanics. Any problems and you phoned their head office in Felixstowe. Then they would arrange to send someone local out – all very time-consuming and problematic.

In the yard at 4 o'clock one morning, my intended trailer had a bald tyre. When I told the night man about it, he replied, 'What the fuck do you want me to do about it?'

Honestly! Couldn't believe it myself!

Perhaps I shouldn't have woken him up. A dodgy tyre is not an emergency, he said. I'd have to wait for the day staff at 7.30am. I went back to bed. It was gone 9am by the time I pulled out, but it was only a short day. When I told the transport manager about the tyre, I also told him my tacho had been in since 3.45am. I can have attitude as well.

The job was pretty easy except for long waits in the docks. Formation queueing around the docks for two or three hours at a time became pretty normal in Southampton during the summer months. And Felixstowe closed down altogether on windy days.

I was doing a lot of writing then, so I didn't really mind too much and made the most of my static time. I was still working on *Diesel Rose*, and having some success with magazine pieces in *Trucking International* and *Truck & Driver*. I often found myself in situations that I could later write about.

One such occasion really brought me out in a sweat. I'd pulled into Thurrock Services one night, just off the M25 at the Dartford Tunnel. My Renault Premium was hooked up to an empty 40' box that

I was loading in Rainham the following morning. The lorry park was full, so I drove around a couple of times before trying my luck in the coach park. That too seemed pretty full of trucks, but one tight slot remained. I was pretty relieved, the alternative would have been a side street in a nearby industrial estate.

To avoid a blind-side reverse (anything to avoid a blind-side reverse!), I drove into the dead-end slot. With a Bulgarian Willi Betz tilt on my nearside and an alcohol tanker on the other, it was a bit tight, but what the heck? I was in. There was a pavement across the end, so I drove up onto it to give my neighbours a bit more room to get out in the morning without disturbing me too much. I had a 10am loading slot, which meant I could have a bit of a lie in.

After a shower, then a cook-up in the cab, I hit the sack and was soon in the land of dreams.

Pop! Bang! Pop! At first it sounded like fireworks. Some idiot having a laugh. Ha ha, now piss off and let me get back to sleep! The noise grew louder and then I heard a lot of shouting. I felt I had to investigate, it was just after 3am. Leaning out of bed, I pulled back the curtains and looked in the nearside mirror.

All I could see was flames! Willi Betz was on fire, his cargo of aerosol cans was exploding, one by one!

I nearly crapped myself! With my heart trying to beat its way out of my chest, I leapt naked into the driving seat and fired up the engine. Turning on the hazards, I whipped back the curtains and whacked her into reverse. I had to get out of there, fast! Looking in the mirror I stopped dead. There was a coach parked right across the back end of the truck!

What now? Abandon truck! Trousers, T-shirt and shoes on in record time, then grabbing my phone and wallet, I was out of there.

Thick black smoke belched out from under my

trailer as I ran for it with my jacket over my head. The tanker driver alongside me was still asleep. I banged on his door yelling 'Fire! Fire!' In seconds he was awake, in his seat and pulling her clear.

Bang! One of Willi's tyres exploded, sending the locking ring flying across the coach park at head height.

Along with half a dozen other drivers, I just stood there, in shock, watching the fire slowly engulf the Bulgarian tilt. The heat was tremendous and my wagon had all but disappeared in the thick acrid smoke.

I was beginning to count my losses – CB, CDs, clothes, bedding, cooker ... – when a cool-headed security guard appeared on the scene. She – yes for it was a she – got onto the coach through the emergency exit, knocked off the handbrake, and yelled at us useless men to push the bloody thing clear. Which we duly did. With the obstruction now out of the way, everybody turned to look at me as if to say 'Well?'

My trailer's flashing hazard lights were only just visible through the dense toxic smoke. But the cab, up on the pavement, seemed clear.

I talked myself into it. Where else would I sleep tonight? How else would I get home? And what about my CDs?

Gritting my teeth, with my jacket back over my head, I sprinted through the smoke and leapt up into the cab. It was so bloody hot! The engine was still running so I slapped her into reverse and stuck my head out of the window.

The other drivers waved me out, screaming at me to be quick. They didn't have to tell me, I was already thinking I'd made a very bad decision. As the cab drew level with the burning wreck the intense heat flushed my face. Then another tyre exploded, sending the drivers scurrying back behind the coach. Now that the tanker had gone I could reverse out in an arc, taking the cab away from the flames.

The first fire engines arrived just as I pulled clear. I drove onto the exit road and collapsed over the steering wheel. That was close, bloody close!

A fireman, knocking on the door, helped me focus.

'There's smoke coming out of your container's back doors!' he shouted. 'What are you carrying?'

When I told him she was empty, he opened up the doors. The handles were so hot he had to use gloves. Smoke poured out, the heat had burnt off the paint from the inside of the container! On the outside she was burnt, black and blistered. All the lights down that side had been burnt off as had the MOT plate and the twist lock handles. The rear lights on the tractor unit were a molten mass of plastic, and the plastic wheel nut covers had all melted off. The trailer tyres had turned a milky white.

Back at the scene, the firemen had doused the flames. The burnt out wreck sat smouldering on its wheel rims. The ribs of the tilt and the skeleton of the cab were all that was left.

The Bulgarian driver stood watching the embers, he was probably wondering what Willi was going to say.

'Chauffeur?' I enquired, pointing to the wreck.

'Ya,' he muttered, expecting a little sympathy.

'Well thanks a fucking bunch for waking me up, pal! Hope I can do the same for you one day!'

I knew he probably didn't understand a word I said, but it felt good to let go. And you know what? I've never driven nose-first into a parking place since.

I later wrote up this terrifying incident and it was published in *Trucking Magazine*.

22

Viva the Revolution

I'VE always had an interest in the politics of the day. As a young teenager I joined the CND and went on an Aldermaston march and several anti-Vietnam War demonstrations. On Boxing Day you'd find me in our town centre, giving out character assessments to the local fox hunt, gathered prior to their search for a fox to rip to pieces. What fun! Merry Christmas!

My dad was a labour voter and a solid supporter of the unions. If you were working-class, he'd say, they were the only ones who ever gave a shit about you. You could never argue with the boss about pay and conditions, you'd just get labelled a troublemaker and face the sack. You needed the shop steward to do it on your behalf. Just like I had at Salbstein's in the 80s.

In big disputes, the media always came out on the boss's side. Union leaders were labelled 'commies' and vilified. Murdoch owned the press and, as we were to find out many years later, had most of the Tories in his pocket as well.

It was about this time that I saw an article about STD – Socialist Truck Drivers. That sounds like me, I thought. A meeting with the founder, Rachel Webb, in Brighton, had me believing that I wasn't alone in this big bad world. There were others that held a similar mindset.

The aim of the STD was to recruit truckers to the party and fight for their rights – better pay and conditions, improved parking facilities, maybe even a bit of respect from their employers and the public.

It wasn't easy getting truckers to sign up for the revolution. A visit to TruckFest in Peterborough found that most truckers supported our aims and ambitions but ...

Trying to get commitment from drivers that lived all over the country and only had a day or two off every week, wasn't easy.

The Socialist Party, of whom the STD was an offshoot, didn't help with their backward-looking attitude. I attended one of their meetings in Brighton, expecting to spend the evening slagging off the government and plotting a way forward. Instead most of the evening was spent slagging of the Socialist Workers Party (SWP), who had broken away from the Socialist Party and was now deemed an enemy. The Tories got off lightly. Their constant use of the term 'comrade' was outdated and off-putting, the need to modernise and move into today's world was obvious.

I hung around for a few months, writing a piece about Willi Betz undercutting the UK haulage business (Blame Willi, not his underpaid drivers). This appeared in *The Morning Star*. Another article about the STD was published in *Commercial Motor*.

But we got nowhere pretty fast. Recruitment flatlined and the in-fighting within the Socialist Party was embarrassing. Small parties often disintegrate in this way. The Green Party did the same in the 90s – after winning lots of council seats, they turned on themselves and imploded. Their battle cry became 'I'm greener than you'. The Socialists seemed to be 'I'm more radical than you'. The media love such squabbles.

I drifted away from the STD, but I don't think anyone noticed.

Securicor morphed into DHL and all the trucks went from turquoise to yellow. Well I suppose anything is better than turquoise. Sorry, is that colour-ism?

They stepped up a grade or two with the motors. New trucks seemed to appear every week. I was given a new MAN for a few days. Great! So when I saw it was automatic, and had no clutch either, I sought help from the office. Nope, none of them had any experience or knowledge of the gearbox either. Just read the manual, they said, and hurry up, time's getting on. Reading the manual never helps until you've been driving the thing for a few days, it's all gobbledygook until then. Well, apart from when she ran away from me, down Birdlip Hill, in Gloucester, I managed to suss it all out.

You didn't have to dip the oil on these modern hi-tech trucks. You just pressed a button and a fairy in the engine block did it for you. And a goblin checks the water level and the elf lets you know if your screen washer is low. Can't have us truckers getting our hands dirty now, can we? And to think I used to do all this manually. And trucks no longer carry spare wheels. Got a blow-out? Make a call and someone else comes out and changes it for you. Now that is progress, should have happened years ago.

On one particular truck, the oil fairy wasn't responding. Just a black line showed up on the dash. DHL continued with the tradition of no garage or mechanics on site. With 30 or so trucks pulling out most mornings, who'd have thought they would ever be needed?

The office told me to phone Felixstowe, who cursed the previous driver for letting it get so low. Must have been an agency driver.

'Bung in five litres,' said the man who knew.

When it made no difference it was, 'Bung in another five!' And when that didn't show he thought I'd better take her to the local MAN dealer.

On the way it soon became clear what the problem was. First I smelled the clutch, acidic and burning. Then red lights and alarm bells began ringing as I pulled into the dealer's yard. He confirmed that the engine had been overfilled with oil. The gauge was showing full, not empty!

Big problem, lots of damage, lots of time off the road. The DHL mechanic whose advice I'd followed got one hell of a bollocking, and so did the last driver who had the truck – an employed driver, not an agency guy.

But me, I was smelling of roses. I was so glad I'd phoned them that knew.

Maintenance certainly suffered by not having any in-house mechanics. Little things were too much hassle to detect, especially near the end of the week. Hanging around for a guy with a spanner to turn up, meant a lot of rushing around to catch up.

Old trailers could be a problem. Apart from being a rough ride, they'd often jam up. You carried a 20' box in the middle of your 40' skeletal trailer. To move it to the rear, for loading or offloading, you shunted the box backwards.

To do this I'd detach the red airline, which brings on the trailer brakes. This is then attached to a pin release valve. This does what it says and enables me to reverse the tractor unit, shortening the trailer like a trombone. Thus placing the container at the rear.

At one particular job, in London and on a Friday afternoon, the trailer was a real pig to close up. The regular greasing of all moving parts must have been missed the last service or three. The pins needed some persuasion from my club hammer before they'd retract. Even then I had to snatch backwards and forwards before I got any movement at all.

Once on the bay, loading began. They'd all had a fag break while waiting for me to sort myself out. In less than 20 minutes, 24 tons of minerals, destined for Dubai, was loaded and the doors sealed up.

To shift the box back into the middle, all I had to do was ease forward with the trailer brakes on, and stretch the trailer open again.

The wheels span, but nothing else happened. Burning rubber made my nose twitch. The weight, all on the back-end, was lifting the drive wheels like a see-saw. I revved and snatched a few times. Nothing. The trailer was jammed solid.

I had to vacate the bay, others were waiting to load. I released the trailer brakes and slowly drove forward. In this position, with all the weight on the back, it was a pretty precarious move. The small yard necessitated a tight right-hand turn and I feared for the tyres as they screeched on the turn.

Once I'd got her straight, I swapped the red airline again and did a few gentle tugs, then a few violent snatches. All I did was leave a few miles of rubber on the tarmac.

I tried everything I could think of. Raise the tractor axle, dump the tractor's air, then the trailer's air. Lower suspension, lift suspension. First gear pull away, second gear, then even third and reverse. The guys in the yard arranged a tug of war with a forklift truck, but that failed miserably.

The problem was a mixture of bad maintenance and the sheer weight on the arse-end. The wet tarmac did us no favours either.

Other trucks came and went, their drivers all coming over and asking, 'Have you tried …?'

I resorted to phoning for help. Felixstowe had no answers, they couldn't suggest anything I hadn't already tried. They offered to send a mechanic, but what could he do? And he was at least a couple of hours away. And it was Friday for God's sake!

After an hour or so of desperation, my saviour finally arrived in the shape of an elderly owner-driver. He ambled over and asked if I had a piece of string. Eh? I thought perhaps he was going to strangle me with it to put me out of my misery.

I supplied the string and watched in wonder as he leaned under the trailer and pushed in the shunt button. This releases the brakes on the trailer, so that shunters who can't be bothered to attach an airline can move the trailer around the yard.

The guy then tied the piece of string around the button and when he told me to drive forward, I suddenly got the picture! The guy was a genius!

As I moved away he ran alongside the wagon. I wellied it faster and faster. When he could no longer keep up, he yanked on the string, the shunt button came out, the trailer brakes slammed on, and with a mighty crack and a bang the container hurled up the trailer and slammed into the stops. I very nearly went through the screen as we screeched to a halt.

I leapt out, and on seeing the box sitting in the required position, punched the air with joy! Yes! Oh yes! I shook the old boy's hand, but I really felt like hugging him.

Stuck on the M25, several hours later, I kept asking myself, 'Why didn't I think of that?'

But the truth is, no matter how long you've been driving, or doing a particular job, you can always pick up a few tips and learn something new.

After I'd been on for over a year, the drivers

stopped stereotyping me. A few even requested to the office that I had their trucks when they took their holidays.

Although to me, some of them were just playing at being truck drivers. The crap that hung in their windscreens marked them out as wannabes. Flags, spears, footie scarves and nameplates. I always removed such rubbish, even if I only had the motor for a day. It would all go into a locker. I've never forgotten the girl in the bobble hat!

I refused to drive one truck. It stunk of piss! The driver was probably one of those morons that pisses in a bottle at night, and then throws it out the window the following morning. But he must have dropped the bloody bottle because the stench was just awful. DHL had the cab steam-cleaned and the driver picked up the tab.

One day, heading back to the yard after a long week away, I'd stopped at some traffic lights, and was just about to pull away on the green when, in the mirror I spotted a young kid, maybe 13 or 14, running across the road towards the trailer. I delayed releasing the handbrake and watched in horror as he ducked underneath the trailer, then reappeared out the other side! His mates on the pavement all cheered, it seemed like a game to them. When I yelled at them, questioning their IQs, they gesticulated wildly and told me to fuck off!

This had happened to me before at Lucking's. If things always happen in threes, maybe I'll squash the next idiot!

Work started dropping off. Still with the agency, ADR, I did a few weeks for Debin's Transport and then a week here and there for Southampton Container Services. Eventually I got a proper job, on the cards, with Maritime. Within a few weeks of starting I was put on a new Topline Scania. What a truck! So much space and boy did she pull!

Maritime were a lot more efficient than DHL. A better, friendlier transport office, and they had their own mechanics!

In early 2003, Jo's sister, Margaret, died after battling cancer for many years. For Jo's mum, Kathleen, it was particularly hard. Margaret was her second daughter to pass this way.

At Margaret's funeral, in Barnsley, Jo complained of a shooting pain down her leg. She began to acquire a limp. Back home the doctor's diagnosis was a trapped nerve, sciatica. It went on for several weeks, getting worse all the time. Poor Jo was in agony.

Eventually she asked for a second opinion, and a tumour was discovered. Inoperable. When Jo asked how long she had the reply was 'less than 12 months'.

The world collapsed around us. Jo refused the offer of chemo, it wouldn't cure, just delay.

I have to confess the following few weeks are just a blur. Jo, so brave, held court daily as friends and relatives from all over the country came to say their goodbyes. She was in great pain, but cheerful and accepting of her fate. Some days the house was so full of well-wishers that there was almost a party atmosphere. She did all she could to comfort those that found the situation so stressful. Which was all of us. Kathleen, losing her third child in this way. Kayleigh losing her beloved Granma. Daniel, in whose body Jo's kidney would live on. And Dylan and Justine, simply distraught that they couldn't do anything to help their mother.

And me. Jo was my lover, my wife and my best friend for 30 years. We'd done so much together and there was still so much left to do. We thought we had the time, but we found out the hard way that tomorrow might not always be there.

A passing clergyman called, Jo said thanks but no thanks. We have never believed in heaven and hell or the afterlife. And Jo never wavered from that. Her life was ending, her time was over. I only hope I can be as brave when it's my turn.

Jo organised her own funeral. She wanted to be cremated in a bamboo coffin, with a humanist ceremony.

She choose her music with much thought. 'Flower Duet' as we gathered for the service, Nigel Kennedy's 'Summer' when we paused for thought and reflections, and Bob Marley's 'Three Little Birds' as we leave. She wanted us all to have a smile on our faces when we left.

The last few weeks of Jo's life were heavily supported with morphine. In her final days at the hospice, her room was often so full that it was standing room only.

THE end finally came on 31 July 2003, just six weeks after she was first diagnosed. She was just 49 years old.

23

Coast to Coast

LIFE changed forever. We all, as a family, revalued our lives. Things that had once seemed important and central to our lives no longer seemed to matter. We had always lived to work, a very British syndrome. You should only ever work to live.

Time passes slowly when you're living in a vacuum. I did a bit of work here and there, containers again through ADR, but my heart wasn't in it.

Dan and Dylan just bumbled on in a dumbed-down sort of way. And Kayleigh grew up very quickly.

Daniel broke out first. He booked himself a six-month snowboarding break in Colorado, USA. We all cheered and said 'Great! Go for it!'

As his plans evolved, he invited me over to join him for a few weeks. Great idea, but snowboarding wasn't really my thing. We played with the idea for a while and it became: maybe hire a car and drive around. This evolved to: fly to Florida, hire a car and drive to Colorado. Yes! A few more beers and this evolved to: maybe carry on driving to San Francisco and make it coast to coast!

On the trip I could take in New Orleans, the Grand Canyon, the Giant Sequoias and climb over the Rocky Mountains. Yes! The idea was born!

No way were Dylan and Kayleigh going to be left behind, so the only way to do the trip was in an RV, a large camper van. We invited two friends, Dean Ellis and his 11-year-old daughter, Quai, to join us.

Jo would have loved all the excitement of plotting and planning as the dream slowly began to become reality. She was with us in spirit on the trip, every step of the way.

I elected myself to do all of the driving for two reasons. One, I loved the challenge of driving all the way across America, and two, like most truck drivers, I'm a lousy passenger. We planned the trip to be a few weeks ahead of the main tourist season.

When we picked up Harvey the RV in Orlando, we immediately headed off to Daytona Beach. Here we filled a bottle with sand and dipped our toes in the Atlantic. Just 28 days later we did the same in the Pacific Ocean, just north of San Francisco, having driven 5,000 miles across the USA.

On the way we spent Dylan's birthday alongside the Mississippi in New Orleans and hunted for diamonds in Arkansas, before meeting up with Dan in Breckenridge, Colorado. After a couple of days trying to snowboard, we took him with us to the Grand Canyon.

Wow! Standing on the edge of this huge chasm is life changing. Believe me. So humbling, it's hard to take in. Goosebumps and hair-raising, this is pure nature in the raw. A mile deep and ten miles across, the multi-coloured strata at the top is millions of years old. At the bottom it's billions of years. The life time of the human race? Just a speck on the canyon's edge. See the Grand Canyon and know your place in the universe!

We saw Californian condors gliding on the thermals, one of the rarest birds on the planet. Camped on the canyon's edge, around a campfire, we listened to coyotes calling in the distance.

I just love wilderness and America is full of it. Huge National Parks and remote campsites miles from any civilisation. Taking the RV off-road, up logging tracks and dirt-trails, we saw real wilderness. In the Valley of the Gods we could have been on Mars: huge red monoliths and rugged mountains in all directions, 360 degrees, like a huge crater. We only saw one other vehicle the whole day we were there.

Dan travelled with us for a few days before returning to Breckenridge. We carried on and visited the craziest place I've ever been to in my life. Las Vegas!

No matter how much you see on the TV and in the media, you'll never be prepared for Las Vegas. It's like a Disney World for grown-ups, a psychedelic acid trip. It's not until you leave and reflect back on it that your mind can actually cope.

The RV park was so big we needed a map to find our reserved slot. Then we walked the 'Strip', Las Vegas Boulevard, which is more than three miles long and lined with casinos and palaces, wedding chapels and steak houses. Flashing neon lights and speakers outside every establishment try to lure you in. 'Spend your money here!' they yell.

The pavements are packed, people jostle shoulder to shoulder. At road junctions people-jams back up, waiting for the chance to dash across.

On the road, four lanes of traffic crawl bumper to bumper. Stretched limos and Humvees jostle for space with pink Cadillacs and huge 4WDs. Through the open windows you could see parties going on inside. The traffic fumes, the noise and the visuals seemed to numb the brain as we shuffled along.

All the while, side-shows are going on in front of the big casinos. At the Treasure Island Hotel we saw two fully rigged man-of-war ships battle it out in a two-acre lagoon. Cutlasses clashed and canyons fired.

Then, after a few rock songs by some sexy female pirates, one ship actually sank beneath the waves! This show was repeated on the hour every hour.

Every tourist attraction in the world seemed to be replicated here. Big Ben was there alongside the Eiffel Tower and the Pyramids. The Statue of Liberty stood facing Nelson's Column. We visited Venice's Grand Canal on the second floor of the Venetian Hotel. Gondola rides were on offer. It claimed to be the biggest hotel in the world. Well it would have to be, wouldn't it?

This is probably why the majority of Americans don't have passports. Why risk dodgy tummies with all that strange food? Why put up with all those nasty foreigners? Everywhere worth visiting was right here in Vegas. And the weather's guaranteed, it's always hot in the middle of a desert!

On the street we formed a human caterpillar, holding onto each other's clothing, to try and stay together in the crush. The noise and the fumes were unbearable. Our noses became blocked, our eyes wept.

Elvis Presley came by in a cape. He was nodding and saying 'Howdy' to everyone. A bride, with an entourage of bridesmaids and wedding guests risked their lives running through the traffic. The vile abuse they got from the honking drivers didn't quite fit the image of this fun-loving place.

The brain became overloaded, you couldn't take it all in. But then, as we made our way back to the RV, we came upon the Bellagio Hotel. Out front, emanating from an eight-acre lake, a thousand fountains shot coloured plumes of water hundreds of feet in the air, swaying and dancing to the soaring vibrato tones of Pavarotti. Brilliant choreography, following the tempo with a flourish, colours changing through the whole spectrum. Fantastic! Repeated every half hour, 24/7.

Vegas blew me away. The girls, along with Dylan, loved it. But I think, having just seen one of the greatest natural wonders of the world, the Grand Canyon, Vegas seemed a little pretentious to me. I was glad of the experience – been there, done that – but it wasn't for me.

After leaving Vegas we crossed Death Valley, the hottest place in America. Venturing out from the RV's air-conditioned sanctuary, you could feel your skin tingling from the heat.

It cooled a little as we climbed up the Rocky Mountains and into the Sequoia National Forest. We paid $10 for access to 1,250,000 acres of wilderness for 30 days. I pay more than that back home just to park a truck in a shitty service station for one bloody night!

It was here we met up with the biggest living thing on the planet. A Sequoiadendron giganteum called General Sherman. When the Romans were invading Britain, the General was just a sapling. Now at 275' tall, he is the dog's bollocks of the tree world. I just stood and stared. He was huge! A circumference of 103', equating to 52,000 cubic feet of timber.

The saddest thing is the fact that the early settlers, in the late 1800s, cut down many much larger ones. One, estimated to be well over 3,000 years old, with a 100' diameter, took five men three weeks to cut down with handsaws.

We stayed over one night in a truckstop and I chatted with a few truck drivers and, of course, ended up sat in a truck or two. Yes, the trucks look the business, and tramping around the states for weeks on end sounds pretty good. But in reality the trucks are pretty crap to drive. Nightmare blind-spots, lots of tiny gauges and a huge gearstick that you have to catch before you can make a gear change. Their pay is as bad as ours. They're usually paid per mile, so if you're not moving, you're not earning, and the status of the trucker is pretty much the same as in the UK. Bottom of the pile!

After travelling through 13 states we ended the trip in San Francisco, 28 days and 5,000 miles after leaving Daytona Beach. We collected sand and dipped our toes into the Pacific Ocean, on the aptly named Dillon Beach. Trip complete, coast to coast.

We all, as a family, benefited greatly from the trip. For me, spending time with the boys and Kayleigh, after years of working away for weeks at a stretch, helped me to adjust to the new order of things.

I kept a diary on the trip and later wrote it up as a travelogue. I had 25 copies printed and gave them to friends and relatives. The kids got their copy so they would be able to reminisce with their grandkids in their old age.

A few weeks after our return, I started working again. Through ADR, I did a bit of container work out of Southampton and Portsmouth. But I wasn't enjoying nights away any more, and I found the weeks long and tedious.

While doing a week's holiday relief for Southampton Container Services (SCS), delays at the docks meant I ran out of time shortly after collecting my box. There was no lorry park in the docks, but you could often get away with parking up outside the café. But not that night.

At about 11pm I was rudely awoken by a fat, spotty security guard, telling me to pull out of the docks or else.

'I'm out of time,' I pleaded, through bleary eyes. 'My tacho ...'

'Don't give me that,' he retorted. 'All you bloody drivers say that!'

My reply might well have been a bit short and to the point. Us bloody drivers kept him in a job. And we needed our beauty sleep.

After my final two-word reply, I wound up the window, closed the curtain and tried to get back to sleep. But it's not easy when you've been wound up.

I tossed and turned until about 1am, when I heard a strange knocking sound, and a few deep whispers.

Clambering into the passenger seat, I wound down the window and stuck my head out. Spotty fat boy was knelt alongside the wheel, bolting on a wheel clamp. Behind him, stood with arms folded, and giving me the glare, was his shaven-headed minder.

'Needs two of you, does it?' I taunted.

'I warned you …' he started.

After giving him a character assessment and a warning about the state of his karma, I retreated to my bed and spent the rest of the night cursing the guard, the docks, the world and all his mates.

First thing in the morning, after kicking the bloody clamp, I phoned my boss. He wasn't happy. He called the security guard who told him he'd release me for the princely sum of £90. He had little choice. His wife would be down with a cheque, in about an hour.

I sat waiting in the café, ribbed endlessly by other drivers. We all enjoy a laugh over others misfortunes.

An old boy, Bobby, sat down with his tea. He soon became my best friend.

'Reckon we could slip that off,' he said, in a West Country accent. 'Looks a bit slack to me.'

So we both went back out and had a good look at the offending clamp. It was a single triangular plate, fastened with chains around the wheel. Slack chains.

On Bobby's instructions, I climbed in and fired her up. With the steering on full left-hand lock, one chain, with the help of a bar, slipped free. On full right-hand lock, the other chain just dropped off.

'Yes!' A cheer went up from a few onlooking truckers. Two of them stepped forward and, picking up the clamp, threw it clean over the fence and into some brambles.

I followed everyone's advice, and after profusely thanking Bobby, I got out of there, pronto. As I pulled away from the docks, an oncoming car, flashing its lights, turned out to be the boss's wife. I pulled over and told her of the great escape. She laughed as she tore up the cheque.

I was only with SCS for the week and on the last day I was again in the docks. And guess what? As I pulled out through security, it was spotty fat boy who stepped forward to take my gate pass from my outstretched hand.

He didn't look up as he snatched it from me. But I couldn't let it go, could I?

'Where's your fucking clamp now, eh? You fat spotty bastard!' I sneered loudly.

Our eyes met and I saw his face change as the recognition set in.

'Fuck you!' I yelled as I accelerated away from the dock gate.

I spent the next month or so running boxes out of Portsmouth. On the odd occasion that I ran into Southampton Docks, I wore my shades and woolly hat, but I never saw him again. I can only assume he'd disappeared up his own arse!

There was still a shortage of drivers around Southampton and I ended up back at DHL. They were so desperate for drivers that they sent representatives to Poland to recruit them and bring them over. Their own drivers were outraged, but were told they were a lot cheaper than agency drivers.

On one of the jobs for DHL, I turned up early with an empty 20' box, at a semi-derelict warehouse in East London. As I swung around in the yard, a guy came out and told me off for being early. Telling me to pull out of the yard and wait in the road, he said he'd call me in when he was ready. He was a big black guy and not at all friendly like.

I sat with the kettle on, keeping an eye in the mirror for Grumpy to give me the nod. A car, a black BMW, came shooting up the road and screeched into the yard. Dust and dirt flying everywhere. Five minutes later Grumpy waved and called me back in.

'Got a loo?' I asked, as I shortened the trailer, moving the box to the back.

'Nope,' was the curt reply.

Sat in the cab, I felt the lurch as a car, I assumed the BMW, was driven into the box. Then some banging as it was secured. Grumpy brought out the paper work, in a sealed brown envelope, and off I went back to Southampton.

Now I don't know about you, but I was a bit suspicious about the whole thing. Something wasn't right.

Back at Southampton I dropped the trailer in the yard for someone else to run into the docks. My next box, for Manchester, was waiting. Going into the transport office, I voiced my concerns to the manager.

'What an imagination,' he laughed. 'Still, make a good story for you, wont it?'

He also told me it was the fourth container in six weeks from that warehouse. All shipped off to Nigeria.

Eventually I gave up the big wage and got work locally, driving for Sandy Bruce. They had a place just behind Ford Prison, near Littlehampton. UK distance work, a couple of nights out each week. Nice people to work for.

Two ex-Lucking's drivers and an old Bailey Boy also drove for them. Great for reminiscing. We'd debate for hours who was the biggest bastard! Gordon or Bigshit?

The Mercedes Actros I was given was the mega-cab and a treat to drive. I also did a bit on a wagon and drag, polystyrene seed trays and tiles to garden centres and DIY stores.

But for me, driving was no longer what it was. At the start of a long European trip, I'd have a buzz about me. Anticipation. Plotting and planning the route, overcoming any problems and obstacles, then finding the tip, getting a reload and heading off back home. And on returning to the yard, I'd have a sense of achievement, job well done. Home safe and sound.

But all that was gone. Now it was just a job. I hated Mondays, and looked forward to the weekend just like everyone else. I was restless, I could see a rut, and I was in it.

At home, Dylan had moved out and now it was just Dan and me. Dan was working long hours as a site manager, and we only seemed to come together at weekends. He had his life and I had mine.

24

Life on the Cut

THEN a crossroads moment. One afternoon in the summer of 2005, on a dilapidated industrial estate just outside Birmingham, I crossed a canal bridge. And there, coming along towards me was a narrowboat. Chugging along, at peace with the world. I was able to stop the truck, and watched as he passed underneath me and went on his way. Such a calm, peaceful scene in an industrial wasteland.

Pow! An image of travelling the UK on the canals hit me. Yes! No motorways, no deadlines, no destination! Carry everything you need. Leave behind everything you don't.

In that instant, on that canal bridge, I regained my buzz. My mind raced with plans and dreams. At the first newspaper shop, I stopped and bought a *Waterways World* magazine. My night out was spent pouring over its pages. Especially the 'Boats for Sale' columns.

Back home, Dan thought it was great and gave me lots of encouragement. A few weeks later I hired a canal boat from Guildford for a week, just to see what living aboard meant.

Of course Dylan and Kayleigh came along and we had a great time, except for getting stuck on a weir! I found the 57'-long hire boat tiny and compartmentalised. Only 2 metres wide – that's why they're called 'narrowboats' – and, apart from a small dining area, the rest of the boat was all beds and storage. But I'd read that 'live-aboards' used their space more wisely. I decided to go for it.

When I first returned from our USA tour with my diary, I needed to get it edited before I sent it off to the printers. Step forward Charlotte Sullivan, a friend of a friend. A music teacher living in Warnham, West Sussex, she helped me get the manuscript to the publishers in a decent readable form. She also encouraged me to follow my dream of living on a narrowboat, and even accompanied me on my trips to marinas, looking for my new home.

If you want to live on a narrowboat, unless you are a continuous cruiser, you have to have an official residential mooring. But continuous cruisers can only stay in the same area for a short space of time before moving on. Therefore if you have a job or a car, you can't really be a continuous cruiser.

Residential moorings are very hard to come by, it meant you needed to find a mooring before you bought a boat. Or buy a boat already on a residential mooring.

Charley and I grew closer. She'd left her home in Toronto, Canada, in her early 20s and travelled around Europe before settling in the UK. She sang on cruise ships and fronted a jazz band before studying to become a music teacher in her early 40s, eventually becoming head of music at the Oakwood School in Horley, Surrey. She also belonged to an amateur drama group and I watched her performances in *Cabaret* and *Oliver*. Great voice!

We went to a few gigs together. BB King was just amazing, and Eric Clapton blew us away at the Albert Hall. Although we listened to most musical genres, the blues was the music that did it for us.

We checked out a few boats at marinas around the

country, but without a mooring, I couldn't really make a decision. I also wanted to travel around the country so I needed to be able to cope with locks on my own. Single-manning a boat is not too difficult, I was told, as long as it wasn't too big. A 57-footer seemed the optimum size. I could go anywhere in the country as most locks were at least that size. A 60- or 70-footer couldn't cross the Pennines as the locks were too small, and they could also be difficult when tying up or manoeuvring in strong winds. Apart from the Guildford trip, all my knowledge came from books and advice from other boaters at the marinas we visited.

An ad in a canal mag had me and Charley rushing up to Northolt, in Middlesex. A boat for sale on a residential mooring!

The Paddington Arm of the Grand Union Canal passes under the A40, in north-west London. Here, the residential moorings at High Line Yachting ran along the private side of the canal. At mooring 30, we had a half-kilometre walk from the car park to reach *Norsk*, the 57' narrowboat we had come to view.

She was beautiful, but alongside her was *Talisker*, a 70-footer, also up for sale.

There was no comparison. *Norsk* was nice, but *Talisker* was amazing. Completely refitted with reclaimed oak, she was spacious and airy. Powered by a 1953 Lister JP3 engine in its own engine room, she looked huge. And she came with her own residential mooring!

I knew as soon as I saw her, she was the one. I had to have her. The owner assured me that he'd single-manned her all around the country.

'Just take your time,' he said. 'Never rush. Life on the canal is lived at a much slower pace to the rest of the world.' Sounded just perfect.

The deal was done. She was mine.

A few weeks later I moved on board. Waking up that first morning, listening to the blackbirds singing, the Canada geese fighting and the seagulls screaming abuse at one another, I just knew I'd found my Utopia.

The mooring itself was the length of the boat and maybe 30' deep. Bordered by trees, it was private and secure. Electric gates, opened by key fob, protected us from the outside world. It was hard to believe that we were so close to the centre of London. It was a little oasis.

Canal-building in the UK first began in the late 1700s, and one of the first linked St Helen's coalmines to the port of Liverpool. Over the next 50 years or so, nearly 2,500 miles were manually dug out, linking all the industrial cities together. Barges were towed along the tow path by horses. The men who drove them took their families along, all living in tiny boatman cabins on the rear of their boats. These were the first long-distance drivers.

But with the first steam trains coming through in the early 1800s, they were doomed to a relatively short time in history. Steam-driven boats appeared in the late 1880s, getting rid of the horse and allowing the boat to pull a 'butty', like a wagon and drag. Diesel engines then followed, but improved roads and an expanding railway network finally killed off the big carrying boat companies.

Nowadays most of the surviving canal network, more than 2,000 miles of it, is mainly used for recreation and pleasure. Although I still get working boats coming by me on the Grand Union almost daily, canal carriers are on the decline. It's so much cheaper by truck, unfortunately.

Although I planned to take the summer off, I

phoned around a few local agencies and discovered that there was plenty of work available. Mostly supermarket and store deliveries. Not my favourite type of work, but I needed to earn a wage to pay the mooring fees, and the waterways licence, along with the insurance and the fuel. Living in paradise doesn't come cheap. Those agencies I spoke to wanted me to sign up there and then. A good sign.

I began planning my next adventure. This year was going to be the third anniversary of Jo's passing. Jo's mum, Kathleen, her sister Heather and a nephew or two lived in Barnsley, South Yorkshire. The Dearne and Dove Canal that actually ran through Barnsley was no longer navigable, and the Sheffield and Tinsley Canal had a length limit of 60'. So the closest I could get with Talisker would be the Chesterfield Canal.

Plans were drawn up to meet up with all the family in Chesterfield on Jo's day. I reckoned on taking at least a month to get there, it was more than 250 miles as the canal flowed.

I'd start off on my own for a week or so, then Dylan would travel up and join me for the last run in. After the get-together, Kayleigh would join the crew for a week or two before going back home. And then Charley would come aboard for the final three or four weeks, running along the Thames back to my home moorings at Northolt.

So at the end of June 2006, me and *Talisker* left the mooring and headed south down the Paddington Arm. Just over two hours later we joined the Grand Union Canal at Bull's Bridge and turned north.

Although the previous owner had told me otherwise, several boat-owners on the mooring thought I was taking a bit of a risk going single-handed. One even said I was a 'bloody fool'. Too late now. When you're chasing a dream, you only hear what you want to hear.

Travelling along 'the cut' was the perfect way to see England from a completely different angle. Passing through farmyards, country estates and pretty villages. It's great for bird-watchers, and the banks are just full of wild flowers, bees and butterflies.

The speed limit on canals is only 3–4mph, a quick walking pace. This is just perfect. Slows you down, you can't hurry anywhere. Gives you plenty of time to take in your surroundings. Mooring up late evenings, I'd sit out, watching the world go by. Maybe have a pint or two and a meal in a waterside pub, then go to sleep in my cabin, listening to the water lapping gently on the hull.

When I passed under the M25, I yelled abuse and gesticulated wildly at the noisy, smelly traffic, carving through the countryside. I could still hear it and smell it several miles away.

Locks were a new concept for me. As with all instruction manuals, they mean nothing until you've been hands-on a couple of times. Paddles? Pounds? And what's a stop-lock?

With locks, the perfect scenario was as follows: It's a sunny day, but not too hot. You arrive at the lock and it's set in your favour. There is no queue and the gates are open. You drive in, tie the boat off, then climb out of the lock and shut the gates using a balance beam. Then you open the paddles with a winding handle, called a windlass, and this lets the water in. Or out, depending on whether you're going uphill or down. Then you open the gates and drive out. Maybe 25 minutes at most.

The worst scenario is this: It's pissing down and the wind is driving into your face. You're going uphill and the lock gates are shut with the lock full of water.

So first you tie the boat up in front of the lock, using at least two ropes and mooring pins. Then you open the paddles on the lock to drain out the water. Now you'll see if you've tied the boat up properly. The water rushing out can flush the boat away as the pins pull out of the soft mud. This usually only happens once. Running down the bank trying to get your boat back teaches you to tie off more securely.

After closing up the paddles and opening the gates, you go back to the boat, untie her and drive her in. Then you have to climb up onto the roof and scramble out of the lock before tying her off. You then shut the gates and flood the lock by opening up the paddles on the other gate. Once full, open the gates, climb back aboard and drive out. This often takes up to an hour or more and leaves you knackered, soaked to the skin and pulling over to put the kettle on.

Locks are a very dangerous place to work, it's so easy to get complacent. Falling in can kill you, as your boat crushes you against the lock walls. Descending a lock can sometimes see the rear of the boat catching on a concealed concrete sill. The front end continues going down and if you can't close the paddles quickly enough, you'll flood right out. Some locks have two sets of paddles. Open them in the wrong order and a torrent of water will pour through your front doors like Niagara Falls and sink you. All these are things that are more likely to happen to single-manned boats, especially when you're tired, cold and wet. You have to stay aware, pay attention and expect the unexpected. Much like driving a truck really.

On good days I could pass through eight to ten locks. Most of the early locks were doubles, taking two boats abreast, so if I ran with another boat, their crew usually did all the work – taking pity on a poor single-manner like me.

I ran with a German family for a couple of days. Their grandchildren had bikes and rode on ahead to set the locks. We'd arrive and drive straight in. The kids would then do all the gates and winding for us before riding ahead to the next lock. We passed through 18 locks in one day.

Engine problems forced me to spend a few days at Braunston sorting it out. A lot of wiring had to be replaced and a new alternator fitted.

Everywhere I went people oohed and aahed about my 1953 Lister engine. A marine legend, they said. But I have to confess, I'd never heard of them before. But then I didn't know *Talisker* was a whiskey either.

But I found the engine to be a pain. The controls were also 1953, you pulled on a rod to engage the gearbox, and turned a small wheel to raise the revs. She responded slowly, so that when I slammed on the brakes – i.e., put her in reverse – it took a while to slow down. I hit a few docks and bridges, putting scars down her side.

After picking up Dylan near Leicester, we joined the River Trent at Nottingham. It was a big river and so different to a canal. Once we'd left the city behind, it became open to the elements, crossing a flat featureless countryside. And after Cromwell Lock at Newark it was tidal.

This was a completely different ball game. A flat bottom boat in a tidal river was not easy to control. The wind could so easily dictate your movements. The tide had to be just right, you needed to run against the tide to keep control.

We waited at Cromwell Lock, along with several other boats, for the tide to turn. The rain began and the wind picked up as we finished our tea and prepared for the off. Lifejackets on and away we went in a flotilla of small craft. We aimed to get to Torksey before mooring up again. There we'd have to wait for the next tide. It was just 16 miles away, but it took us nearly four hours.

In places the river was probably 200 or 300 yards wide. You had to stay in the middle of the river as it snaked around the bends. A couple of times we got too close to the shingle banks and ran aground. A quick reversing manoeuvre was needed to get us off and back out into mid-stream.

My hands were numb from the cold. The wind whistling up the river made the steering bloody heavy and my arms ached.

'Behind you!' yelled Dylan, stood on the front and pointing back down river.

I turned and saw a ship, a small coaster, bearing down on us. I say small, but he was a bloody sight bigger than us. He was obviously going faster than us and we had to move to one side to let him pass. The wake from the ship as it sped by threatened to roll us over. A frightening experience.

We over-nighted at Torksey, sharing a small jetty with other boats alongside a pub. A ploughman's and a pint was all I could manage. I was exhausted and knew there was more of the same to come tomorrow.

We set off again at first light. The tide has turned overnight and had left us with less than 5 hours to get to West Stockwith. Everything on the boat was straining at the max. The engine was at top revs, flat out, and I was knackered, flat out. My map was now just a sodden mess, very little use at all. The river had no recognisable landmarks. Along the bank, tall kilometre poles slowly passed by, giving us an indication of how slow our progress was.

We needed to exit the River Trent at West Stockwith Lock, where we'd pick up the Chesterfield Canal. I'd phoned the lock-keeper before we set off and he introduced me to the nightmare that was Stockwith.

He asked me how much experience I had. That sent alarm bells ringing straight away. None whatsoever, I replied. He told me of the dangers that lay ahead.

By the time we got there, the tide should be on the turn. This would make it easier for us with little or no current. But if we got it wrong, we could end up being swept up the Trent and out into the Humber Estuary. No return.

The lock was on a sharp right-hand bend in the river. If I turned in too soon, what current there was could turn the boat around and we'd soon be going past backwards! If I turned too late, I'd miss the lock completely and not get a second chance. Both scenarios meant the Humber would be our final destination.

But fear not, the lock-keeper had a strategy. He'd stand in a visible position on the side of the open lock. He would watch for our approach and as we drew level with the dock gates, he would wave his arms above his head to indicate when to start the turn. He assured me that this was standard practice, he did it several times a day. I was going to ask him what his success rate was, but decided against it.

We took nearly four hours to travel the 15 miles from Torksey, by which time the tide had pretty much turned. The lock-keeper phoned a couple of times telling us to hurry up.

When the lock gates finally came into view, I felt a relief similar to that of coming off Pamplona with Teapot in the blizzard. We'd made it in one piece.

But we weren't safe and sound yet. Now we had to get off this bloody river and into the lock. Thankfully the gates were huge, the lock was big enough for at least half a dozen boats.

Our guiding light stood waiting for us. Stay in the middle, he'd said, and turn at right angles when the signal came. Which I did, gunning the Lister for all she was worth. But as soon as she was broadside to the current, she began drifting past the lock entrance sideways. I swung on the tiller with all my might, yelling at her to come around. By the time I'd straightened her back up, we were a couple of hundred yards past the bloody lock. With the power full on, we fought the oncoming tide and struggled back towards the lock gates. Progress was pitiful, at times we seemed to be standing still.

I sussed that we now needed to go right past the gates before trying again. I think the lock-keeper agreed, he kept waving me on. This time I was closer to the bank, if I couldn't get her in this time, I might have no option but to run her aground.

The lock-keeper started jumping up and down, waving me in. I went for it, gunning her to the point of blowing her up. She was screaming out for mercy as her nose turned and she crept slowly towards safety. She started to turn too quickly, as I swung on the tiller we swept into the lock, bouncing of the walls and clipping the huge gates. But I didn't care, we were in! Off the bloody river! That was close. Bloody close! I wouldn't want to try that again.

As the lock filled and rose us up to the Chesterfield Canal, I thanked the lock-keeper, without whom ... He laughed and told us he had only lost a couple of boats in all the years he'd been there. If we'd had shit out, he said, the Humber Lifeboat would have come to the rescue. That's all right then.

Over the next few days we ran on down the canal, and at Worksop, alongside a waterside pub, we all met up for a family get-together on Jo's anniversary. Justine, Daniel and Kayleigh had travelled up by car from home and, after a good meal, everybody piled onto *Talisker* and we did a trip up and down the canal. A good day out was had by all.

When we pulled back out, heading for home, Kayleigh joined us for a week. Getting out of Stockwith Lock was no problem as we just slipped out of the lock and into the river, going with the flow. The trip back down the Trent was far more enjoyable, with sunshine and the satisfaction that we'd done the hardest part of the trip.

Dylan jumped ship at Nottingham and Justine picked up Kayleigh a couple of days later at Leicester. I ran solo for a few days, enjoying the solitude but missing the extra pair of hands at the locks.

Charley met up with me at Braunston where we joined the Oxford Canal, slowly heading south down to the River Thames.

We cruised the Thames for a couple of weeks, passing Windsor Castle and Henley. At Reading the rock festival was on. We tied up just behind the main stage and 'enjoyed' the wind-distorted music right into the night. The next day we met up with Dan for a meal in a riverside hotel.

Leaving the Thames at Weybridge, we joined the River Wey and ran down to Guildford. After a few days cruising, Charley went home and I made my way back up to the Thames, heading for my home mooring.

Coming off the Thames at Brentford I joined the Grand Union Canal. Between Brentford and Bull's Bridge, at the start of the Paddington Arm, lay 10 locks in a six-mile stretch. Every bloody one of them was set against me. It took me nearly ten hours of hard slog to get through, and stopped me from being too sentimental about the end of my great adventure.

But I did get a great sense of satisfaction from completing the trip unscathed. It had taken me 11 weeks to cover 583 miles and pass through 326 locks. I learnt so much, a lot of it the hard way. And I'd bonded with *Talisker*, my beautiful boat and new home.

25

Wagon and Dragging in London

BACK at the mooring, I relaxed for a week or two before signing up to an agency. There were plenty of permanent jobs around but I didn't want to commit to anything that had me working weekends or nights away. Being with an agency meant I decided when I wanted to work and when I didn't. As long as I gave them sufficient notice, it was never a problem. I'd never be rich, but I was certainly happy.

My relationship with Charley had taken off. We'd spend a weekend on the boat, then the following weekend we'd be at her place in Warnham, just outside Horsham in West Sussex, my old home town. Being a teacher meant she had half-term breaks and summer holidays. When Charley had time off, so did I.

Driving Logistics, based in Eastcote, just up the road from my boatyard, were a small friendly agency whose promise of lots of work actually materialised.

I started working for the John Lewis Partnership, pulling out of Park Royal, in West London. I'd run a wagon and drag down to Peter Jones in Sloane Square three or four times a day. Cages, roll on, roll off. Nice steady pace and a great bunch of people at both ends.

They were the most diverse and multi-cultured company I've ever worked for – the flags of all their employees were displayed at the main entrance. With a workforce of just 120, there were 36 different nationalities, all working together happily and successfully.

Workplace discussions, over the news of the day, were so enlightening. Abdula, a Muslim from Algeria, told me that there were no Jewish victims in the 9/11 attack. He strongly believed that it was all an American plot to discredit Islam. All Jewish workers in the towers had been warned the day before not to go to work, he claimed. No matter how many holes there were in this theory, he was adamant it was true.

William, who came over with his parents as a child from Jamaica, tried to convince me that the world was only 10,000 years old, and created by his God. That made my huge fossil collection seem a bit redundant.

Being born into a Catholic family, I was born a sinner, expected to pray for forgiveness every day of my life. To a kid, this was like being told off for something you never did. By the time I'd reached my early teens, I was already a soap box atheist.

Both Abdula and William found it difficult to understand someone like me, who didn't believe in a creator. How could I not believe in an afterlife? What was the point of living if there was no rewarding paradise at the end?

I'd counter by saying that you have just one life, make the most of it. If you don't fear death, a natural conclusion to your life, you can achieve happiness here on earth. But sorry, once you're dead, that's it. End of.

Our debates and discussions were all friendly and very respectful. I admire people with faith, the discipline and structure it brings to their lives must be very rewarding.

Although sometimes, like a naughty schoolboy, I'd find myself playing devil's advocate. Walk on water?

Seventy-six virgins? What if you came back as a flea on a cat's bottom?

The John Lewis Partnership is unique in that the workers own and run the company. Every employee, and there are more than 76,000 of them, is an equal partner in the firm. There are no shareholders, all profits are shared among the partners every year in the form of a bonus. Most years this bonus equates to between 15 and 20 per cent of their salary.

The partnership owns hotels, caravans and lodges all over the country at holiday locations. All subsidised for family holidays. Partners also get a hefty discount on purchases from JLP and Waitrose stores, along with free entry to National Trust properties, Kew Gardens and a host of other leisure activities, plus discount cinema and theatre tickets. And they've even got a boat if you fancy a bit of sailing in the Solent.

They have their own constitution and really seem to look after their workers. A lot of their drivers have come up from their shops and warehouses. Free training and encouragement for all. The managers of most departments have also risen through the ranks. Then once you've served 25 years, and a lot of partners do, you are entitled to 26 weeks paid leave! That's six months! Most of them seem to take off on world travel and cruises. A great reward for all those years of service.

The sickness schemes are very generous and, after long periods of illness, partners can return to just a few hours a day to ease themselves back to work.

To me, the John Lewis Partnership is the nearest we're ever going to get to socialism in the workplace. No unions, but workers' forums with elected officials seem to do a better, more personal job than they ever could.

Occasionally I'd do a week or two for Marks & Spencer or Sainsbury's, but the bulk of Driver Logistics' work was for John Lewis. Variations included runs to Oxford Street and trunking around their various Midland depots.

I'd been on for JLP over a year when work began to slacken off. I took the summer off and Charley and I did a tour of Spain and Portugal in an old VW T4 van, with a bed in the back and an awning that plugged on the side. A real busman's holiday.

We visited Odemira and I recalled the days of being a Bailey Boy, one-hitting it back to the ferry for peanuts and abuse. It's shameful that I put up with it, but different times, different situations. I'd come a long way since then.

The downturn began to bite after the bankers were caught with their trousers down. I was only getting a couple of days work a week for months on end.

Making the most of my time off, I self-published *Diesel Rose*, my first novel, and had more magazine articles published. I started making progress with my next novel, *Charlie's Place*. Nothing to do with my Charley, I began writing it several years ago, it was a bit of a slow-burner. The story is about a crook and his dysfunctional family who run a truckstop. It's sex, drugs and rock 'n' roll all the way.

I needed to earn more to pay all the mooring fees and licences. So I took a permanent job with G&P Transport. George Webb, who ran the small firm was a really nice guy. He ran a few eight-legger tippers on a tarmac contract and a couple of artic tippers running sand and stone to local ready-mix plants. At the interview, I made it clear that I didn't work weekends. George accepted that, although I got a bit of banter from his other drivers because they all had to.

The work was easy and the truck, a Volvo artic, was almost perfect for the job. I say almost because she had an automatic gearbox and I wasn't a fan. I think this comes from running down mountains – no way would I want the truck to decide what gear to go down in, or when to change down. I'd managed OK with them at DHL and Sandy Bruce, but found it difficult to get the best out of them. Reversing could be very awkward. With a manual box you tend to slip the clutch a bit to ease back gently onto the loading bay. With an auto, you just go back until you hit. Nothing gentle about it at all.

At one of the ready-mix plants, I had to reverse up a slope to tip the sand in a bin. If you reversed too slow, you'd come to a halt, then she'd run back down the slope. Reverse too fast and you'd smash into the concrete wall and burst your trailer's air bags. Which I did the first week. Thankfully George was forgiving and my suggestion of laying a railway sleeper just in front of the bin to stop the truck solved the problem.

The yard was also very small and I had to jack-knife the wagon tightly around to line her up to the slope. At full lock, when the weight of the trailer lifted off the pin, she'd come to a halt, rock wildly back and forth, then take off again. All very unpredictable. Maybe I should read the manuals. Maybe I was just a dinosaur.

There was a large contingent of Polish drivers working out of the plants. They were a great bunch of guys, working hard and sending money back to their families at home.

One day, one of them told me he hadn't shown his daily break on his tacho. He asked me what to do. I advised him just to write on the back of his tacho disk that he'd accidentally used the wrong mode. Then he said, 'What, on all of them?'

It transpired he'd worked for the firm for more than six months and had never once shown a break. He would just put his card in in the morning and take it out when he finished, not changing modes at all. He'd been driving for years, but he said nobody bothered with tachos in Poland. And he had every tacho for the past six months in his cab! More than 21 is a very expensive offence.

When he asked me if he should throw them all away. I said no! It was his boss's fault for not making sure he handed them in. He should also have checked he was running straight. I suggested he hand them all in now, and let his boss throw them away.

I was dismayed by the underlying racism within the Polish drivers. One day I asked one what he thought of England.

'Great place to live,' he replied, 'but you let too many blacks in!'

Another time one of them had moved from Ealing to Richmond.

'What's the new place like?' I asked.

'Very nice,' he replied. 'There's not a single n★★★★★ living in the street.'

My lecture on diversity and the benefits of multi-culturalism got hoots of derision.

It saddens me when, even if you've only known people for a few months or so, a hidden side suddenly leaps out and smacks you in the face. You question your ability to tell friend from foe.

On a run around the M25 one day, heading for a quarry, a passing truck was being tailgated by a wagon and drag. I always feel truck drivers should know better, we're the professionals, right? In the best of conditions, anything less than an artic's length between you and the guy in front is plain suicidal. At 50mph, you're travelling at 22 metres per second!

Just looking down at your radio or checking your mirrors could see you ploughing into oblivion if an emergency occurs.

As the offending truck drew level with my cab, I looked over and shook my head in disbelief at his stupidity. He replied with a single finger, I responded with a knuckle shuffle. His response was to back off, great I thought. Then he pulled in behind me and I became his next victim. He was just feet behind me, flashing his lights wildly. I think I may have upset him. The truth often hurts.

I tried to ignore him. A touch of the brakes from me and we'd both be in trouble. As I carried on round the M25 he stayed with me, bobbing in and out, arm waving out of his window.

When I pulled off the motorway at the St Albans slip road he followed. It was a two-lane motorway slip. As I slowed, he swung around me and went past. When his drag was level with my cab, he came in! Of course I was expecting it, but I still had to brake hard and swing over to avoid a collision. We came to a halt blocking both lanes.

The driver appeared at the side of my cab, and I'd swear steam was coming out of his ears. Jumping up and down on the spot, he screamed at me to get out of the cab so he could kill me! I declined the offer. He tried my door. Locked! I wasn't that stupid. His face resembled a beetroot, the veins in his neck were throbbing! Then as I dialled 999, requesting urgent back up, he climbed up on the step and, with one punch, smashed the wing mirror. Pressing his face to my window, his spittle splashed the glass as he raved on with his desire to send me to hell.

A queue of cars watched as he continued jumping up and down. He'd lost it completely. He suddenly leapt up onto the bumper and, screaming unintelligibly, head-butted the screen before grabbing the wiper blades and bending them double.

After repeating his threats to prematurely end my life, several times over, he returned to his cab and drove off. I drove to the first lay-by and waited for the cops.

Now you may remember my own road rage incident at Hammersmith, with the lady I made cry. Yes, I behaved like an arsehole that day, and I regretted it immediately. But this guy was a psycho, no doubt about it. The thought of him driving around in a 40-ton truck, with that mindset, was enough for me to want him off the road forever.

The police took photos of the damage and filed a report. It dragged on for months with phone calls coming and going, and the case being passed from one police officer to the next. They found out who he was, not difficult I would have thought, and in a letter, 'requested' that he attended a police station. When he didn't respond, they wrote to him again. And again. The case fizzled out and I eventually lost interest, as nobody seemed too bothered at all. What I saw as criminal damage and threats to kill, the police treated as a domestic spat! Perhaps if he kills someone next time, they might do more than write a bloody letter.

In July 2010, after 12 months or so working for George, I took the summer off, and Charley and I went backpacking around Cuba. Travelling around by taxis and buses, we spent the first and last night in a hotel, the rest of the time we lodged in 'casa particulars'. These were people's homes – a bit like our B&Bs, but a lot more personal and friendly.

The taxis were mostly old American Cadillacs and Oldsmobiles from the 50s and we could travel miles around the island in them for just a few quid.

There were a lot of old American trucks on the road as well. Mainly old Macs, and Kenworths imported via South America to avoid sanctions. Some old Russian trucks still survive from their time as Cuba's protectors.

Visiting the Bay of Pigs, where those damn Yankees got their arses kicked, I sat in the same chair that Fidel Castro had sat in as he oversaw the defeat of the American-backed invasion force. I posed for pictures with statues of Che Guevara, hero of the revolution.

There was music everywhere we went – in every bar and on every beach. In Havana we caught the latest line-up of the Buena Vista Social Club. Brilliant night.

Even now, with Obama as president, Cuba is still subject to crippling trade sanctions from the USA. The blockade has gone on for 50 years now. The Americans can point at Cuba's struggling economy as a failing of socialism, but they have to accept some responsibility. It could all change tomorrow, with a little bit of goodwill from both sides.

Back home, after a short break, I returned to working for John Lewis. This time out of their place in Acton. The work was pretty constant. I'd do a week of running an artic into their Oxford Street store, then the next week I'd be off to their depot in Blakelands, Milton Keynes, twice a day. Then it would be the Sloane Square run with a wagon and drag all week. All very constant and all at a steady pace, no rushing or pressure. Nice people to work for.

The biggest problem, running in and out of the centre of London every day was, without doubt, cyclists. Every year an average of ten cyclists die on London's roads. And another ten are seriously injured every week. More than half of those that die are in collision with a truck or bus. Shocking statistics. But from my own observations, the only surprising thing is that it's not a lot more.

I could so easily add to these statistics on a daily basis. Very few cyclists have had any formal training since doing their cycling proficiency badge at primary school. Us truck drivers are trained, retrained and trained again all through our career, yet the finger of blame is always firmly pointed at us. We are the baddies, tearing around in juggernauts, polluting the atmosphere, careless and uncaring about other road users.

On the wagon and drag, I have six indicators down the side. When they flash on and off, it means I'm turning. Simple, we all know that. But it's never enough. Professional drivers look ahead, constantly assessing and reassessing conditions ahead as they arise. Eyes like radar, sweeping the road, ahead, behind and alongside.

Cyclists on the other hand, ride heads down, arse up, completely unaware of the traffic around them and the possibility of any conflict. Roads are a very dangerous environment, you have to be aware at all times.

One day, in Notting Hill Gate, a bus in front of me slowed and indicated a left turn. As I backed off, a cyclist undertook me and, in spite of clear indicators from the bus, began to undertake him. He was halfway up the side of the bus before he realised he was about to be squashed into the kerbside railings. Forced to stop, he began screaming abuse and banging violently on the bus windows.

I sounded my horn, and when he looked over I made it abundantly clear who I thought was the tosser.

Another day, a two-wheeled maniac undertook me when I'd stopped at a zebra crossing. A mother and

251

her two kids had to jump for their lives. He couldn't have stopped even if he'd wanted to, because he had a cup of Costa coffee in his hand. He wouldn't have heard the abuse the woman shouted at him either, because he had a giant set of headphones on.

And is it not illegal to use a phone on a bike? And what about umbrellas? If that's not against the law it bloody well should be. I see these things daily. They think they're immune to everything and above the law. When they are involved in an accident, it's always someone else's fault. Especially if that someone else is a truck.

I've had them clinging to the back of my wagon, leaning against the side at traffic lights, and one actually held on to my wing mirror. Anything seems better that putting their feet down. And do they not realise how annoying it is when they ride two abreast, in deep conservation, as half of London queues behind them?

'Boris Bikes' have added to the problem. More than 200 have been written off since the start of the project. And who can forget the TV footage of Boris himself, promoting his bikes by riding to work on one. Despite the fact that several news teams were following him, he casually jumped two red lights and rode the wrong way down a one-way street. What a role model!

If I had a pound for every cyclist I saw jump a red light, I'd double my wages. And whatever happened to bells? The only warning you're likely to get these days is a yelled curse and a finger or two.

As the law stands there is nothing to stop a blind man or a toddler with stabilisers on her bike from cycling down the Marylebone Road in the rush hour.

Training for cyclists is the way forward to cutting casualties. Tax and insurance might be a good idea too. I'm surprised that the penny-pinchers in government haven't already noticed another cash cow in waiting.

I cope with most bad car drivers by accepting that I'm the professional and they're not. That's not arrogance, it's a survival mode. It stops me getting uptight. Drivers on school runs, on their way to the supermarket or racing to catch a plane have other things on their minds apart from lane discipline or correct indicating at roundabouts. It's up to me to make allowances for their slack behaviour. The odd beep or shaking of head in disbelief is all they're likely to get these days.

There are now more than 24 million driving licence holders in the UK, of which more than half a million are HGV drivers. From my own experience, UK drivers are the best in Europe. The best of a bad bunch maybe, but once you've driven in Greece, Portugal or Italy, believe me, you can appreciate the calm saneness of a Monday morning jam along Marylebone Road. Of the 120,000 deaths on Europe's roads each year, 'only' about 2,000 occur in the UK.

All professional road-users in London have their own agenda. The worst drivers I come across are the minicabs. They're identified by a round green permit stuck in their rear windows. I call them 'tosser stickers'. I don't know what training they get, if any, but the standard of their driving is simply appalling. They're chancers and bullies.

I find the black cab drivers selfish and rude. If I hesitate for one millisecond on a green light, somewhere behind me a cab will sound his horn. Yet every day I see cabs dropping off and picking up their punters at traffic lights! And they never seem to hurry or even be embarrassed about it. And why do they never pull over to the kerb to pick up their punters? A metre or so away is the best you can hope for. So you're often forced to wait while they load up or disgorge.

Reversing a wagon and drag into the back entrance of Peter Jones can be a bit slow and laborious, depending on how many cars there are on the double yellow lines. I always imagine that most car drivers held up by my delicate shunting and manoeuvring are admiring my skills and appreciating the difficulties involved. But the constant beeping of cabs, further back in the queue, does tend to grate.

Somewhere in the Highway Code there's a bit about 'care and consideration for other road users'. But like indicators and politeness, this is rarely seen among London cabbies. And why do they run on side-lights at night? Shit ones at that. You even see them on the motorways, middle of the night, on bloody side-lights. Professional drivers call these 'parking lights'.

Trucks can be a pain as well. With so much construction work now going on in the capital, eight-legger tippers are everywhere. They always seem to be in a rush and are the worst tailgaters you're ever likely to see. And they will never let you out from a side street, something that really pisses me off. If you're in a slow-moving jam, why not let someone in or out, especially if that someone is a fellow trucker?

Traffic jams are great levellers. It doesn't matter if you're driving a Ferrari, a truck or a Trannie, we are all in the same boat. No matter how much money you've got in the bank or how little, we're all stuck in it together.

During the late summer months, the wealthy residents of Knightsbridge take to the roads to show off their toys. Chrome Porches and gold Ferraris compete with Lamborghinis and Bugattis as they fly by Harrods before roaring down Sloane Street and around the Square in an opulent show of wealth and ego. Most of the cars are on Arabic plates, meaning they've been shipped over. A recent TV report said the gold Ferrari was worth £250,000, and only driven a few times a year. In times of food banks and welfare cuts, this all seems a bit off-balance to me.

But the *Big Issue* seller outside Peter Jones thinks they're just great. Makes his day to see them all parade past.

John Lewis is very high-profile. Customer relations are everything. The shop floor staff bend over backwards to please: nothing is too much trouble, the customer is God.

Their shopping would be carried out to their car, sat on double yellows across the loading bay. You just smiled as they apologised profusely for holding you up. The success of the business showed them that the customer wanted to be loved, treated well and appreciated. And always be right (even when they were obviously not).

On one of my wagon and drag runs, I ran through roadworks on Cromwell Road, West London. Four times a day, five days a week, for more than three weeks. Three lanes down to two, plenty of warning, nearside lane closed.

I'd always take the middle lane early, otherwise traffic would never let you out in time. A silver Ford didn't seem to notice, however, and stayed alongside me seemingly oblivious to the closed lane. When at last she noticed (yes sorry, for it was a she), she just indicated and drove into the side of my truck.

I was pretty shocked. I knew she was there, but I expected her to back off, not to hit me. I cleared the road, pulling over into the roadworks.

I wasn't uptight at all – accidents happen, everyone makes mistakes now and then. Nobody was hurt, but the young driver did seem a bit shaken. She wouldn't look me in the eye and kept saying to me. 'Well I did indicate!'

'Yes,' I replied. 'You did, but then you drove into me.'

My passenger step was trashed and she had a crumpled wing and had lost a wing mirror.

'Well you never used it anyway,' I laughed, trying to lighten things up. But she insisted I was in the wrong for not getting out of her way.

Her companion got out of the car but never said a word in her friend's defence. In fact she didn't speak at all.

I pointed out that she had her lane and I had mine. She had come into my lane. I didn't hit her, she hit me! I never got cross at all. Exasperated yes, but not uptight.

It was a company car and she seemed concerned about what her boss would say. Forms were duly filled out and we both went on our way. The transport office were sympathetic and the damage easily repaired. And then I forgot all about it.

Over a month later, the transport manager showed me a letter that the management had received, regarding the accident. The woman had written to head office complaining about me. She claimed I'd driven into her, got arsey and frightened her. Nothing was further from the truth. She also claimed I'd said I hadn't seen her. Ridiculous, of course I'd seen her, and even if I hadn't, I certainly wouldn't have said so.

She retold my joke about her wing mirror, making my transport manager laugh. But she obviously found it life-changing and hurtful. The truth often hurts.

Her description of the accident bore little resemblance to the truth. If she made the same claims on her insurance form, surely it could be seen as fraud.

She ended the long rambling letter by stating how she'd been a John Lewis shopper all her life, but now would never enter their stores again!

While most people who read the letter agreed that it was probably an arse-saving letter for the benefit of her employers, I was not allowed a right to reply. It did piss me off a bit. If she could write to my boss, why couldn't I write to hers?

Don't make a fuss, let it go. That was the John Lewis way. But looking back now, I'm OK with that.

I did a week or two of night runs. Not starting until 5pm, then running up to Ollerton, north of Nottingham. Then a trailer change before returning via Blakelands. I'm not really into night shifts, but now and then I enjoyed a change. It also made me feel like a real trucker again. It was good knowing I could still do the big hits if I had to.

26

Supermarket Crap

LIFE on *Talisker* was living the dream. On Charley's half-terms, I'd take time off and we'd go away on a trip up the Oxford Canal or down to the Thames. Waking up in a different place every day was like being back up the road. But at a leisurely 4–5mph, your whole life seemed to slow right down. Days are longer when you take your time. Good for the mind and soul.

For bird-watching it was just perfect living on the canal. In most gardens, a count of 18 to 20 different birds throughout the year would be considered pretty good. Here on my mooring, it's more than 40. Of course living on water helps – kingfishers, moorhens and cormorants aren't often seen in gardens!

John Lewis spoiled me a bit. I started to believe that the agency driver was an important part of the haulage industry, a valued part of the set-up. But I was soon brought down to earth when I did a few weeks for Sainsbury's, in Greenford. It was a huge operation with hundreds of trucks pulling in and out 24 hours a day.

You had to be assessed before you were taken on. This was, of course, in my own time and unpaid. Their transport operation was run by Wincanton, with my agency, Driver Logistics, having an office on the premises.

Out on the road with my assessor, I was surprised

and dismayed, when he just sat in the passenger seat sending and receiving text messages. Hardly a word passed between us in the 30-minute run around. This showed a complete lack of respect for me, and any respect I should have had for him went right out of the window.

The job itself wasn't much better. The staff who manned the transport office weren't at all friendly. They had to deal with up to 20 different agency drivers every day, so you could maybe understand the lack of a personal touch, but a smile or maybe a response to my cheerful 'good morning' could have made all the difference.

You had to clock on, and then complete a mountain of paperwork before you got the truck. Most forms were just covering Wincanton's arse. You had to state that you were legally qualified to do the job and promise to drive carefully and obey all the rules. Then more paperwork for the truck, check the tail-lift, check the fridge, check the load. And more paperwork at the security gate before you were allowed to leave. Once out on the road it was a piece of piss – but the paperwork was soul destroying.

I stuck with it because the John Lewis work was drying up and I needed to earn. It's not cheap, living on a narrowboat.

But then I got a text message from my agency telling me to arrive at Sainsbury's ten minutes early so I could complete the paperwork *before* I clocked on. Enough was enough. 'Don't send me there again' was my terse reply, and I never went back.

Work was a bit patchy for a while. I even considered seeking a proper job, but I was enjoying the freedom that agency work gave me. Some weekends Charley and I would take *Talisker* up into Little Venice in Central London. A leisurely two-hour trip through golf courses and industrial sites, with an aqueduct over the North Circular Road. We'd stay over and see a band or maybe a musical, then travel back to the mooring on the Sunday. We both loved live music.

I also had plenty of time for my writing, working on my new novel and having several pieces published in *Truck & Driver* magazine.

I did a few days for Lloyd Frasier in Greenford, shifting clothes for H&M and Ann Summers. Mostly shunting between their warehouses or trunking up to Rugby overnight.

Then the agency got the Morrisons contract, out of Feltham. I got a week of 4am starts. Not my best time of day, especially as I rarely left the yard before 6am. A lack of respect for all drivers, not just agency, seemed apparent. I was chastised for failing to ensure my ticks on the daily driver questionnaire were neatly in the box. A young paper jockey in the office curtly told me to do my hi-vis up. And when I started back after 14 days off, I was told that I had to convert it into hours, 'cos it said hours on the form'.

Then one day as I was doing my safety walk-around, first thing in the morning, a woman came out of the office and told me to hurry up as I was late!

What? I couldn't believe it! I snapped back at her. 'No, I'm not late! I was here on time! Everybody else may be running late, but not me! Do you want me to do a proper walk-around, or shall I just tick all the bloody boxes whatever?'

She seemed a bit shocked by my response. She walked away and never spoke to me again. I feared I had become a grumpy old man.

Again, once you got out of the gate, the job was OK. Shop deliveries with cages and a tail-lift. No problem and always plenty of help if required. But the shop-workers always seemed stressed, understaffing seemed to be their main gripe. If I was a shoplifter, Morrisons would be my local.

While doing a run to Oxford for Morrisons, I experienced another road rage incident, and in similar circumstances as my last case. Running through some roadworks on the A40, a truck behind me was so bloody close all I could see was his wing mirrors. I flashed my hazards to try to get him to back off, but got no response. He was probably too close to see them!

Once out of the roadworks he came flying by, and to my surprise it was a Yaris chemical tanker!

I was shocked! One false move by this maniac and there could have been a major incident. Who knows what nasties sloshed about in his tank?

Once upon a time, tanker drivers were the elite among drivers. A dangerous job, well rewarded and a lot of respect shown. But in recent years, standards – along with pay – have dropped. Like most jobs, you pay peanuts, you get monkeys!

I obviously hadn't learnt my lesson from the last incident, because as he drew level with my cab I looked over and tapped my forehead, mouthing the words 'Fucking idiot!'

He responded like the moron he so obviously was. He came straight in on me, forcing me to slam on the brakes and swerve onto the hard shoulder. I came to a halt, crapping myself as his trailer missed my cab by a gnat's breath!

I struggled to comprehend his mindset. Like the maniac on the M25, this guy is still out there, coming soon to a road near you.

Charley and I took off again, touring Ireland. This time with a transit van, kitted out with a bed in the back. What an amazing country. The scenery is so spectacular and every pub, in every town had live music! We loved it.

We followed this up with a two-week tour of Snowdonia and the Lake District. Call them busman's holidays if you like, but it was great to be back to waking up in a different place most days, and I just loved plotting and planning the routes.

I went back on for John Lewis, but they were slowly winding down their Acton depot. A lot of the runs were now going direct from their Magna Park super-warehouse in Milton Keynes. I got the odd day here and there, but I needed more.

My agency called one day and told me of a new contract.

'Suit you, Mick,' he said. 'They're looking for old-school drivers.'

Was this a compliment? I wondered. What's 'old-school'? Professionals that stick to the rules, look after the wagon and get the best MPG? Or drivers that go that extra mile (or hour) to get the job done at all costs? Well done son!

I got a start. On meeting the transport manager I introduced myself and offered my licence and driver card for inspection.

'Later,' he said, handing me some keys. 'We're running late, we'll sort that later.'

We never did. And all those agency forms and arse-covering affidavits that were the bane of my life at Sainsbury's and Morrisons? Never saw one.

This was a bit worrying. Yeah, I know, I whinge on and on about all that paperwork and procedural crap, but take it all away and I must say it makes you feel a bit vulnerable.

The depot, in Southall, ran ten trucks out of an old factory yard. They were double-shifted, running day and night. Other trucks would also run in overnight, leaving trailers for us to deliver on the next day. Cement, soft drinks, crisps – anything really.

Surprisingly for an operation that ran throughout the night, the office opened at 8am and closed at 5pm. So when you turned up at 4am and there was no paperwork, no truck or no load, which happened often, you had to use your initiative or get the transport manager out of bed.

Your initiative meant finding your own delivery notes among a large pile in security, then opening up several trailers to find one that matched your paperwork. If your truck was not back from the night shift, you just took somebody else's wagon.

But it was so much easier just to get the guy out of bed. He was used to it. Let him make the decisions.

The driving was totally different from shop deliveries. I was running curtain-siders up and down the country. Different destinations every day, maximising my driving hours. It was much harder than pushing cages around, but I began to enjoy my driving again. Shop deliveries numb the brain. You just come in, do the job, then go home again.

And I was earning more as well. Life was looking good.

Then one hot sunny day, I was reminded of my mortality. The M25 was running smoothly and I was on my way home. The Renault Premium was a nice drive, comfortable and smooth, easy on my old bones. The auto-box was one of the best I'd driven. After years of slagging off the automatics, I'd finally found one that could cope with reversing and tight U-turns. Is it the trucks that are changing or me?

This is the sort of truck that lets you sit back and enjoy the ride. Although pulling 44 tons, with a limiter set at 52mph, she did tend to back off a bit on the steeper slopes. But on the downhills, I'd let her roll, careful not to let her hit the infringement-bearing heights of 60mph or more.

Anti-clockwise, between junctions 28 and 27 is one such stretch. Long and clear, I let her go, just using the exhauster to hold her back. On the radio, Ken Bruce was doing his pop master's quiz. I'd just answered three in a row when suddenly all hell broke loose.

An almighty bang was quickly followed by the cab shaking violently and the steering snatching hard to the right. A front wheel blowout!

My hands locked onto the steering wheel as instinctively I hit the brakes but, as the wheel snatched even harder, realised it wasn't the thing to do and quickly took my foot off. My left hand pulled down on the wheel and, as my right hand pushed up, my elbows locked.

A quick glance in the mirror saw nothing but smoke and bits of rubber spewing out from below the cab. I pushed back into the seat as the shaking cab threatened to dethrone me. Just about holding her straight and with my heart in my mouth I dabbed at the brakes again. Again the wheel nearly spun from my grip. So I just let the exhauster slow her down as I fought to stay straight and in control. It took me a couple of hundred yards before I finally got her, and me, to the safety of the hard shoulder.

When she eventually came to a halt I whacked on the handbrake, turned off the engine, then collapsed over the steering wheel. My heart was racing and I had pins and needles from the tips of my toes to the top of my head!

That was close! Fucking close! These were the words that ran through my head, over and over again.

A knock on the door. A couple of builders who'd been following me in their van had stopped to see if I was all right. My lips were a bit numb but I managed to thank them for their concerns. They said how lucky I was to get away with it. They didn't need to tell me!

It was a good ten minutes or so before I felt able to phone my boss and tell him the news. Yes, I was OK. And so was the truck.

It took a couple of hours for the tyre guy to arrive and in that time I reflected on my terrifying experience. I frightened myself by playing 'what if?'

What if I'd been overtaking? Or changing lanes, pulling in or pulling out? What if I'd only had one hand on the wheel? What if I'd been having a drink? Eating a banana? Checking the map? I think the outcome could have been rather different.

Had I done the right thing? What if I'd just slapped on the brakes in panic? Perhaps I should have put her in manual and gone down through the gears. But I was just holding on for dear life, trying to keep her straight. It would have taken a crowbar to prize my hands off the steering wheel.

I can't recall any training, CPC or otherwise, that prepares you for such an event. But I do recall hearing tales of trucks having blowouts and going through the centre crash barriers.

The tyre guy arrived and the Highway Agency closed off the nearside lane to allow him to fit a new tyre. Even though the lane had a red flashing lights over it, cars, and a few trucks, stayed in the lane till they could see the cones. Apart from the obvious dangers to all of us stood around the wagon, don't they know it's three points on their licence?

Events like this can change your life. If it's happened to you once, it can happen again. And if it's never happened to you, believe you me, it can. The

cause of my puncture was a bolt lying on the carriageway, probably came off the back of a truck.

So the next time you're rolling down a slope,
having a scratch or unwrapping that Yorkie Bar, why
don't you play 'what if?'

Racism, not only in our industry, but in society as a
whole, still angers me. We're constantly being told
that we, the Brits, are becoming more 'tolerant' of
our increasingly multi-cultured society. I think not.
My experiences sadly have shown little change over
the years.

At a recent tip, a driver asked me which depot I
pulled out of. When I replied 'Southall', he laughed
and said that couldn't be true as I was white and spoke
English. When he saw that I wasn't amused, he confessed he'd never been there.

The forklift drivers tipping us were Polish; this got
him ranting about immigrants pinching 'our' jobs.
Before long he was slagging off the UK and our way
of life, telling me he was checking out a truck driving
job in Canada. He wanted to move out there with all
his family.

But wouldn't that make him an immigrant? I asked.
Wondering what the Canadian truckers would think
of him pinching their jobs?

'It's different,' he replied. 'I'm English.'

Since the year dot, us humans have migrated
around the planet in search of food and security.
Modern-day America and Australia were both built
on immigration.

I remember the Thatcher years, when unemployment was at its worst. Thousands of British workers
went over to a booming Germany seeking work.
Sleeping rough in tents and sheds, they were ripped
off by their paymasters and hated by the German

workforce for undercutting them and stealing their jobs.

But who could blame these hardy folk for trying to pay their rent and feed their families? They even made a TV programme about it, *Auf Wiedersehen, Pet*. A comedy. Ha ha.

But now Britain is the host nation, it's not so funny anymore.

Some are fleeing from wars and terror: wars often fought by the West protecting their interests. Others come from the poorest corners of Europe. When you've got nothing in life but a hungry family, you really want to believe the tales of streets paved with gold.

'When you've got nothing, you've got nothing to lose,' sang Bob Dylan. He was so right.

27

Reflections

THIS year I 'celebrated' 40 years on the road. I can still remember the euphoria when, against all the odds, I passed my test. A miracle really.

And that bastard, Sergeant Major! What would I say to him if we met up today? Probably not a lot. I might even feel sorry for him and his stressed-out way of life. His life peaked in the RAF, it was all downhill from there. Living in the past eats away at your very soul.

My life began with that pink pass slip. My passport to earning a wage and endless travel. For the truck driver, things have changed immensely over the past few decades. When I first started driving, the M1 was just two lanes with no crash barrier. The motorways at night were so empty that trucks would flash and wave if they saw another wagon coming the other way. Now with 'just-in-time' deliveries and the online shopping frenzy, the motorways are often just as busy at night as they are in the day.

The biggest change of all for us truckers has to be the mobile phone. They've changed the world, along with the human race. When they first arrived in our cabs, we all thought they were wonderful. No longer did we have to find a working phone box where you could actually park less than a hundred yards away. Remember those reverse-charge calls?

No longer do we sweat and stress when stuck in a

traffic jam with an urgent load. We just make a call on our phone, and the problem isn't ours any more.

It's great to think we can talk to anyone, anywhere in the world, from inside our cab. But they can also call us, and the job's stress levels rose as they did. We went from phoning in once a day, to being available 24/7.

People have adapted to their phones in a way Darwin would have struggled to understand. On the street everyone seems to have one on show. Either pinned to their ear or in their hand waiting for the next incoming call. There seems to be no kind of etiquette or even any embarrassment at interrupting a conversation or a meeting to take a call. You're the one being rude if you show any sign of annoyance.

Sat in a public loo one day, I listened in on a most interesting conversation about a new conservatory being fitted. As the guy sat for a crap, he had a builder measuring up his house on the other end of the line.

The last time I had a haircut, the guy took two calls, pausing each time and forcing me to listen to his inane conversation regarding his child's progress at drama class.

In warehouses and on loading bays, forklift drivers and loaders seem unable to go for longer than a few minutes in between checking their phones for messages. It really makes you wonder how we ever managed without them.

With all our new technology, transport offices are now run like battle control centres. Your routes are given to you and you're expected to stick to them. And watch out if you don't, because your boss knows exactly where you are every minute of the day. Spies in the sky record your every move for posterity.

While driving for Murfitt's we had to take a diversion one day due to a road closure. Within minutes a message flashed up. 'Why are you going the wrong way?' it demanded. On the Morrisons contract, I'd stopped for a pee and was interrupted mid-flow by a call wanting to know why I'd stopped.

And before you've even returned to the yard, the traffic office will know your MPG, your driving style and how many times you've braked a bit sharpish (although not the reason why). They will even know what colour underpants you've got on. (OK I made that bit up, but it will come, believe me.)

You're then sat down in front of a computer and debriefed by the office idiot who points out everything you've done wrong and explains in a patronising way how to improve your driving skills.

Running down to Spain and Italy all those years ago, part of the buzz was that you chose your own route. Stopping where you wanted to, starting and finishing your day when you chose to. I just loved trying different routes, especially over the mountain passes in the summer, or coastal roads with views out over the Mediterranean. I always preferred the scenic route if time allowed.

Satnav has been around now for more than ten years. How did we ever find our way home from work without it? I laugh myself silly when I hear of trucks and cars ending up in rivers and fields because they've followed them so blindly. A motorist drove down a railway track because her satnav told her to. Professional truck drivers strike bridges because their satnav tells them to drive on. A coach driver in Scotland phoned the police for help when his brain, sorry his satnav, failed. Don't they print maps anymore? Is common sense being bred out of the human race?

My delivery notes now often just state the name of the customer and a postcode. With no help from the spy in the sky, this can cause problems. My stock reply, when asked why I don't have satnav is, 'I've got a map and a brain'.

A bit flippant, yes, and it could be a struggle to reach my retirement without having to obtain one. And then it will probably be just like my laptop. Took me years to get one, then, of course, I wished I'd done it years ago.

Cars can now park themselves, and their lights and windscreen wipers come on automatically when needed. Our trucks communicate with the garage and book themselves in for a service. They're now fitted with lane control warnings, to stop you drifting when you fall asleep, distance control, to stop you before you hit the guy in front, then there's rollover stop and jack-knife stop. If anyone comes within a couple of feet of your moving wagon, alarm bells ring and voices warn you of impending catastrophe.

Is it just me or has all the fun (or should that be responsibility) been taken away from us? Are we just steering wheel attendants now?

I'm bloody glad we don't change punctures any more, but I use to carry a hammer and a screwdriver and was able to sort out most problems that arose when miles away from base. Now I'm not allowed to even change a bloody bulb. Make a call and a 'technician' will come out to sort it. OK I suppose, but a bit of common sense could save a lot of time and expense.

One thing that hasn't changed much is the public's attitude to trucks and their drivers. We're just mad, bad, hell drivers, charging through their towns and cities in huge juggernauts, frightening the locals and polluting the atmosphere.

In the rest of Europe, we're shown respect and appreciation for the job we do.

Truck lobbyists in the UK, who know little about trucks or logistics, rant on and on, demanding curfews, weight limits and outright bans. These people need to be educated in the real facts of life.

They need to be told that everything they own, from the cars they drive, to the food they eat and the houses they live in, every component, brick and sausage, travelled on a truck somewhere on its way into their lives. The plants in their garden, the toilet paper they wipe their bums with and the soap to wash their hands, all delivered by a truck.

The only thing we don't deliver are babies! We would if we could but we're too busy running over cyclists!

The latest cry is for all trucks to be banned from cities in the rush hour. Great! So each truck would be replaced by at least three or four smaller vehicles instead. No way would the traffic flow improve or pollution decrease. And of course haulage prices would have to rise.

If the lobbyists spoke to the industry instead of shouting to the media, I'm sure we could all get together and help improve the situation.

They say the majority of us truckers are now in our 50s and 60s. So what happens when we all retire or die? HGV courses and driving tests are so bloody expensive, few youngsters can afford them. But why would they want to? You can earn better money sitting at a desk or working in a store. No responsibilities, no pressure and no ever-changing rules and regulations to get your head around. Does any other profession get fined for working too hard?

Our wages now are the worst they've ever been. Ten years ago I was taking home nearly £700 a week. Week in, week out. Yes, I was putting the hours in and away all week, but the money was there if you wanted to earn it. Nowadays I'm lucky if I can clear half of that. OK, so the work's a lot easier and often I only do seven- or eight-hour days, but doesn't skill and responsibility count for anything anymore?

I'm proud to tell people I'm a professional truck driver. But the job has changed beyond recognition and I'm looking forward to calling it a day.

I've changed as well. I've already hinted I may be becoming a grumpy old man. I don't put up with things any more. I'm not rude, I hope, I just tend to say things as I see them.

My feelings towards auto-gearboxes have also changed. Remember how I hated them? But they really are the future for trucks and are getting better all the time. I don't tell too many people, but now when I climb up into a truck, I'm pretty pleased when I see it's an automatic. Sitting in traffic jams all day long with a manual box kills my elbow and knees.

Another change for the better is the attitude to health and safety. People take it seriously now, not just paying lip service to the inspectors when they call. I think the fear of litigation from injured workers may have helped it along a bit. Whenever I visit a site and see forklift drivers wearing trainers, it puts me on my guard. But then you get places like CEMEX, where you have to wear full PPI with a hard hat and safety glasses just to drop a bloody trailer in their yard.

I'm able to retire at the end of 2015, but I'll still have to work a few days here and there to be able to put wood on the fire. To be able to carry on working, I had to take my CPC – Certificate of Professional Competence. I had no choice. Somebody somewhere had an idea to make us drivers more competent. Obviously not a truck driver.

My CPC cost me more than £400 out of my own pocket and I had to turn down a week's work to do the course. I had to attend five one-day courses to gain my 35 hours training. To be repeated every five years of my working life.

I could have attended the same course five days in a row. I could, as several of my classmates did, sleep the whole way through the lesson. Others played games on their phones and iPads. There is no test or means to show you learnt anything from it at all. It's an attendance-based qualification.

Now I'm not saying that in my 40 years behind the wheel, I know all there is to know about the job – I don't and I never will. And I did learn quite a bit about the mechanical side of things and the importance of health and safety. All good things to know and understand.

But am I being selfish by not understanding why CPCs were brought in retrospectively? Why didn't they bring in this new legalisation for new drivers coming into the system? Make it part of the test? This is the way most rule changes come in.

We 'oldies' learnt our craft out on the road, we learnt through experience. Good and bad. We all make mistakes, and when you cock up you eventually have to admit it, if only to yourself. It's the way us humans evolve.

You learn to check your handbrake's on before you leave the cab. Because once, you didn't. You learn to double-check your pin for the same reason.

When I got blown over in the wind in Lucking's, it took me years to accept some of the responsibility. I'd witnessed several trucks on their side, and heard on the CB of many others. I should have parked up. But I allowed myself to be bullied by Crouchy, so it was really down to me. It wouldn't happen today.

When I lay the tipper over, I should have realised

the white sand was waterlogged and tipped accordingly. The signs were there. Again it took me a while to see this.

No one showed me how to change a wheel, I learnt the hard way on Bromley High Street. The same way I learnt to rope and sheet, strip a tilt and deal with a front wheel blowout at 60mph! The only way I got down Pamplona that day in the snow was because I had Teapot sitting alongside me telling me what to do. No manual or classroom lesson could have bailed me out.

This is how I, and many others like me, learnt our craft. You can learn in the classroom how things should be, but not how they often are.

I'm not sure that the 'good old days' ever really existed. Working your nuts off, shit pay, employer abuse, you put up with it because you had to. When Gordon called you a c★★★, you laughed or you cried. Your choice. When Bigshit wanted that urgent load back, you ran bent or you collected your P45. Murfitt's sacked us for saying no. Your principles don't pay the mortgage.

Of course most of that has changed now, thankfully, but the driver is still the lowest common denominator in the world of transport.

I don't enjoy driving trucks as much as I used to. Just doing store deliveries and day work is a lot different to doing an eight-day trip to Italy or Norway. But I don't really want to do distance work anymore, I love being back on-board *Talisker* at the end of a busy day. Even when we go off travelling, I'm always glad to get back to my garden and my own bed.

I've finished *Charlie's Place* and it's doing the round of publishers. No luck yet. *Truck & Driver* still publish the occasional piece – usually my rantings and ravings about us poor truckers. I still get a buzz when I see my scribblings in print.

I'm getting old, in body but not in mind. My elbows ache and my knees creak, but I still go out to play. Leonard Cohen, Santana and Beverley Knight in *Memphis*, just some of the gigs we've managed to catch recently.

Last Christmas I asked Charley to marry me. And she said yes!

We're going to grow old together. Plans for the future include a trip over the Canadian Rockies and a visit to Brazil before we're too old to samba. Backpacking of course!

Life is good. This truckin' man is showing no signs of slowing down.